Uses of Blogs

Steve Jones
General Editor

Vol. 38

PETER LANG
New York • Washington, D.C./Baltimore • Bern
Frankfurt am Main • Berlin • Brussels • Vienna • Oxford

Uses of Blogs

Axel Bruns + Joanne Jacobs, EDITORS

PETER LANG
New York • Washington, D.C./Baltimore • Bern
Frankfurt am Main • Berlin • Brussels • Vienna • Oxford

Library of Congress Cataloging-in-Publication Data

Uses of blogs / edited by Axel Bruns, Joanne Jacobs.
p. cm. — (Digital formations; v. 38)
Includes bibliographical references.
1. Weblogs—Social aspects. 2. Weblogs—Political aspects. 3. Weblogs—
Economic aspects. I. Bruns, Axel. II. Jacobs, Joanne. III. Title. IV. Series.
HM567.U74   303.48'33—dc22   2006000642
ISBN 0-8204-8124-6
ISSN 1526-3169

Bibliographic information published by **Die Deutsche Bibliothek**.
Die Deutsche Bibliothek lists this publication in the "Deutsche
Nationalbibliografie"; detailed bibliographic data is available
on the Internet at http://dnb.ddb.de/.

Cover concept by Joanne Jacobs and Axel Bruns
Cover photo by Gavin Winter
Cover design by Lisa Barfield

The paper in this book meets the guidelines for permanence and durability
of the Committee on Production Guidelines for Book Longevity
of the Council of Library Resources.

© 2006 Peter Lang Publishing, Inc., New York
29 Broadway, New York, NY 10006
www.peterlang.com

Printed in the United States of America

Contents

Acknowledgments

Above all, we would like to thank the many contributors to this book, who have provided so many valuable insights into their diverse areas of expertise, as well as the multitude of bloggers who have made this field of research such a rich and exciting environment.

Additionally, Axel would like to thank colleagues and friends who have supported work on this book, especially Terry Flew, Jude Smith, John Hartley, Stuart Cunningham, and Stephen Towers at Queensland University of Technology; as well as Steve Jones, Trebor Scholz, Liz Ferrier, P. David Marshall, and Geert Lovink for valuable advice. As always, many thanks also to the M/C–Media and Culture team (http://www.media-culture.org.au/), and particular gratitude also to my family and friends for their support. Finally, a big thanks also to Jo, especially for bringing her considerable expertise on blogging, and her network of blog experts, into this project.

Joanne would like to thank Axel for taking on the idea of this book over a cup of coffee one day, then doing much of the work in setting up the book proposal with Peter Lang USA, and coordinating information on–and editing of–chapters. Thanks also to the crew at the Brisbane Graduate School of Business at Queensland University of Technology, and at the Australasian Cooperative Research Centre for Interaction Design, for supporting this project. Finally, Joanne would like to thank her mother, Alison, and brother, Andrew, for their constant love and support.

Axel Bruns   http://snurb.info/
Joanne Jacobs   http://joannejacobs.net/

Brisbane, April 2006

Introduction

Axel Bruns & Joanne Jacobs

B logs, it seems, are everywhere. Like few media phenomena, and certainly like no media form since the emergence of the World Wide Web itself, blogs seem to have captured the public imagination. Indeed, the very term "blog" itself was chosen as Merriam-Webster's "word of the year" 2004.[1] Blogs played key roles in the U.S. presidential primaries (even if the world's first mainstream blogger-candidate, Howard Dean, crashed back to earth when the broadcast media joined the party); bloggers were invited to cover the national conventions of both Democrats and Republicans; and blogs also played a significant role in reporting unfolding world events from the London underground to the streets of Iraq, to the shores of Indonesia and Thailand. CNN, BBC, newspapers, and other mainstream media now regularly turn to the blogosphere to gauge public opinion on controversial issues, and this coverage of "what the bloggers are saying" has begun to replace the traditional vox-pop interview with the person in the street.

But to focus on such early achievements of blogging, notable though they may be, is to miss out on much of the variety of opinions, ideas, knowledge, and creativity that can be found in blogs. While certainly a sign of the shifting balance between what bloggers have come to call the MSM—mainstream media—and the average citizen, to discuss only those blogs that debate the news or express political views ignores much of what exists in the wider blogosphere, beyond the pundits and citizen journalists. Similarly, much debate of blogging appears to focus on the so-called "A-list" of established, well-known, and often controversial bloggers while bypassing the vast range of other participants whose engagement only makes the blogosphere possible. Without them, the A-list would be little more than a collection of personal opinion sites; it is only the intercast between blogs and bloggers known and unknown—however uneven in its

> **A-List:**
> a common term used to refer to the best-known (most read, most linked to) bloggers—many of whom blog about the news, or about blogging itself.

power relations—that has turned blogging into the global phenomenon it has become.

Far from being defined by the activities of a Lawrence Lessig, Meg Hourihan, Glenn Reynolds, Esther Dyson, or even Salam Pax, Weblogging is a broad-based movement, not the province of a select few. In 2004, according to a study by the Pew Center, adoption of this emergent tool grew at a rate of 58 percent in the United States alone; 32 million Americans were blog readers by the end of the year, and "7% of the 120 million U.S. adults who use the Internet say they have created a blog or web-based diary. That represents more than 8 million people."[2] While there is some information on who these people are likely to be (male rather than female by a small margin, probably under thirty years of age, relatively well-off, well-educated, and well-connected via broadband), what remains unknown is just how they are using blogs both as readers and as producers of content. Only one fact is certain: not all of them will be engaging in news commentary and political discussions.

(Many) Uses of Blogs

This collection is born out of our need to begin to better identify and understand these uses, to map them and chart their implications for those who engage in them as well as society at large. As Clay Shirky has put it in his discussion of the power law distribution between A-list bloggers and the wider blogging community,

> at some point (probably one we've already passed), weblog technology will be seen as a platform for so many forms of publishing, filtering, aggregation, and syndication that blogging will stop referring to any particularly coherent activity. The term "blog" will fall into the middle distance, as "home page" and "portal" have, words that used to mean some concrete thing, but which were stretched by use past the point of meaning. This will happen when head and tail of the power law distribution become so different that we can't think of J. Random Blogger and Glenn Reynolds of Instapundit as doing the same thing.[3]

We agree that this point has now passed; indeed, we would argue that the fact that the Pew Center study also found only 38 percent of U.S.-based Internet users to have an understanding of the term "blog" might not indicate complete unfamiliarity, but rather also leave open the possibility that these users had encountered blogs under a different guise: as *LiveJournals*, as community discussion boards, as news bulletins, or as creative outlets. Asking users what a "portal" is might generate a similarly low response rate, and yet many are no doubt using portals as part of their everyday Internet diet.

Beyond the basic definition of "blogging" as the reverse-chronological posting of individually authored entries that include the capacity to provide hypertext links and often allow comment-based responses from readers, then, the term "blog" now has little meaning unless a descriptive qualifier can be attached. In the future it is likely that we will come to speak primarily not of blogging *per se*, but of diary blogging, corporate blogging, community blogging, research blogging, and many other specific sub-genres that are variations on the overall blogging theme. Our discussion of blogs, bloggers, and blogging must become more sophisticated; it makes as little sense to discuss the uses of blogs as it does to discuss, say, the uses of television unless we specify clearly what genres and contexts of use we aim to address.

This multiplicity can be a threat, too, as it can prevent an effective understanding and utilization of the technology: because there is now a vast variety of styles of blogs—from online journals to *de facto* news sites, learning tools, knowledge bases, and community support spaces, to name just a few—it is difficult for some disciplines to imagine how blogs might be used for their benefit. Further, and partly because of the mass amateurization of publishing online, blogs have been criticized in some professional sectors as being illegitimate or even dangerously skewed information archives. Yet it is the specific implementation of a blog that determines its value: its operational structures and response mechanisms, as well as the style of writing and method of recording ideas, commentary, and institutionally relevant information, all influence the significance, reputation, and success of a blog. There is a clear need to interrogate the range of blogging styles used by different disciplines and cultural groups and to develop a lexicon to articulate the most effective blogging mechanisms for different contexts. Examples of how blogs are already being used can provide some insight into how they can be further developed for particular interest groups and industry sectors. The use of blogs in generating competitive advantage, and their application as knowledge management tools, is crucial to understanding the relevance of blogs for a range of professional organizations as well as community groups.

Uses of Blogs

In charting some of the current uses made of blogs, then, this book is organized into three distinct sections. Drawing on the experience of blogging pioneers and researchers from a variety of professional and community contexts, the book documents the growth of blogs online and provides a detailed scholarly analysis of successful and powerful blogging uses. We begin by investigating blogging applications in various key industries, such as the highly visible and controversial practice of news and filter blogging as a supplement to mainstream journalism, and the increasing uses of blogs in publishing and business. Our contributors examine the economic benefits of blogs and blogging for fields such as public relations and marketing, and consider their impact on the wider economy. Toward the end of this section, they contemplate emergent uses of blogs in legal and educational contexts, and evaluate the process of engagement with the medium of blogging in these fields.

Section two looks at the social and communal aspects of blogging. From general analyses to very personal accounts of the effects of blogging in particular contexts, chapters in this section explore the value of blogs as a means of social expression and participation and consider how the negotiation of identity online can be realized through the instrument of blogging. Authors in this section provide views on the impact of blogging on scholarly authority and the communication of research in the academy. They also examine the politics of blogging with a focus on traditional party politics, gender politics, and subcultural community participation. Further chapters particularly explore the not always comfortable intersection between blog technology and its users, looking variously at the extent to which such technology might empower or disenfranchise disabled users, at different conceptualizations of blog hosting environments in the South Korean blogosphere, and at how blog technology has spawned a new genre of fictional blogging.

Finally, section three reviews some of the technological and legal possibilities for blogging, such as the expansion of the form from a predominantly text-based mode to one that incorporates audiovisual materials, as well as the legal questions that arise from the blogosphere's complex and multilayered interlinkages that cross global jurisdictions. We present some ideas about the impact and ramifications of blogging and about the future development of blogs and their uses across industrial and social contexts. The book concludes with the contributors' biographical details and a bibliography that draws together some of the key texts and resources used by our authors.

We acknowledge that this bibliography, and indeed our coverage of blog uses itself, is by no means exhaustive; neither can it be. Bloggers and blog researchers may feel that their specific, favorite form of blogging has not been covered in enough depth, or that especially some new and emerging approaches to blogging have not been sufficiently represented in this collection. This is unavoidable, however—even when we first developed this book project in late 2004, some of the more recent uses of blogs (such as corporate "dark blogs," or fictional blogging) had barely emerged to public attention. Fast-paced as development in this field continues to be, studies of blogging must for now remain temporary snapshots of practices in the blogosphere. A new edition of this book a few years from now might present a significantly different collection of articles, using terms as yet uncoined.

Blogs and the Blogosphere

At the same time, while even today some commentators already debate the use of the word "blog" to describe this plethora of online publications, we would suggest that the nature of these publishing efforts still merits use of that term to describe the method of informing an interested public about issues and ideas. Again, in analogy to television, there remain certain shared social, organizational, and technological features that delineate blogs from other forms of online publications (even if the boundaries are blurred).

For example, it is the social networking of blogs and the potential for collaboration that provides a decidedly human dimension to the publishing and publicizing of information. By personalizing content, blogs go beyond a purely informative role and provide a platform for debate, deliberation, and the expression of personal identity in relation to the rest of the (blogging) world. Replacing somewhat outdated email lists and personal Websites as vehicles for exchanging ideas and information, blogs represent for authors an opportunity to reach out and connect with an audience never before accessible to them, while maintaining control over their personal expressive spaces. The blogosphere, understood as the totality of all blogs and the communicative intercast between them through linking, commentary, and trackbacks, is perhaps unique in its structure as a distributed, decentered, fluctuating, *ad hoc* network of individual Websites that interrelate, interact, and (occasionally) intercreate with one another.[4] And it is here, rather than in individual blogs, where the power of blogging is situated; as Hiler writes, "it's not the individual weblog that fascinates me. It's when you tap the collective power of thousands of weblogs that you start to see all sorts of interesting behavior emerge. It's a property of what scientists call complex adaptive systems and it's enabling weblogs as a collective to become more than the sum of its parts."[5]

> **Blogosphere:**
> common term to describe the overall community of blogs and bloggers, which is interlinked through a large number of cross-references between individual blog entries.

In this environment, anyone with access to the network can participate; the barriers to entry are low, and there is no central authority to grant publishing rights or accreditation, nor to prevent bloggers from linking and responding to information or ideas found elsewhere on- or offline. Thus, in a time of redefining our value systems, of competing belief frameworks, and of global threats, bloggers have the chance to question their understanding of issues, engage in discussion, present their ideas, seek out approval for their notions, and grasp some sense of purpose, order, and hope.

As a distributed, broad-based practice of content production, blogging can be seen as a key sign of our times. The industrial, mass media age was dominated by the value chain of the production of physical goods, from producer to distributor to consumer, and perhaps best exemplified by the assembly line invented by Henry Ford (who gave this Fordist model of industrial production its name). This model extended to manufacturing as much as to the media, where audiences were similarly regarded as mass consumers, and the maximization of audience shares was the highest goal.

From Production to Produsage

Today, as the information age replaces the industrial age, the Fordist mass production model has been replaced by one of individuation, personalization, and customization, but this is only a first step: from customization follows interaction, from interaction follows interactivity, and from interactivity follows, in the right setting, intercreativity. This undermines the distinction between commercial producers and distributors on the one side, and consuming, passive audiences on the other; participants in interactive spaces are always more than merely audiences, but instead are users of content. Further, if they become involved in intercreative environments (as bloggers do), they are also active producers of content.

In becoming active publishers, commentators, and discussants, then, bloggers turn into what we can usefully describe as *produsers*—a hybrid of producer and user.[6] All bloggers are both potential users (in the narrow sense of information recipient) as well as potential producers of content, and the blogosphere overall is an environment for the massively distributed, collaborative *produsage* of information and knowledge. This conceptualization advances well beyond Alvin Toffler's famous term "prosumer," which at worst may describe little more than a well-informed consumer who nonetheless remains engaged only passively once a consumption choice has been made, and may never actively engage in the production and expression of new ideas. Shirky similarly argues against the professional consumer:

> **Produsers:**
> users of collaborative environments who engage with content interchangeably in consumptive *and* productive modes (and often in both at virtually the same time): they carry out **produsage**.

> in changing the relations between media and individuals, the Internet does not herald the rise of a powerful consumer. The Internet heralds the disappearance of the consumer altogether, because the Internet destroys the noisy advertiser/silent consumer relationship that the mass media relies [sic] upon. The rise of the Internet undermines the existence of the consumer because it undermines the role of mass media. In the age of the Internet, no one is a passive consumer anymore because everyone is a media outlet.[7]

The media, however, can also be seen as producers of the perception of community, and thus of society at large: media help us understand who "we" are and how we relate to the societies we live in. If, as blogging and other collaborative media phenomena appear to indicate, there is now an ongoing shift from production/consumption-based mass media, which *produce* a vision of society for us to consume as relatively passive audiences, to produsage-based personal media, where users are active *produsers* of a shared understanding of society which is open for others to participate in, to develop and challenge, and thus to continually co-create, then this cannot help but have a profound effect on our future. At worst, it may generate more debate and disagreement, as long-standing values and traditions are questioned; at best, it may offer renewed hope for a more broad-based, democratic involvement of citizens in the issues that matter to them. Understanding the emerging uses of blogs, and the patterns of interaction, intercreation, and produsage that they enable, is an important step in charting the path ahead and in identifying the obstacles and opportunities that we may encounter along the way.

NOTES

1. *BBC News Online*, "'Blog' Picked as Word of the Year," 1 Dec. 2004, http://news.bbc.co.uk/2/hi/technology/4059291.stm (accessed 26 Oct. 2005).

2. Lee Rainie, "The State of Blogging," *Pew Internet & American Life Project*, 2 Jan. 2005, http://www.pewinternet.org/pdfs/PIP_blogging_data.pdf (accessed 5 Jan. 2005), p. 1.

3. Clay Shirky, "Power Laws, Weblogs, and Inequality," *Clay Shirky's Writings about the Internet: Economics & Culture, Media & Community, Open Source*, 2 Oct. 2003, http://www.shirky.com/writings/powerlaw_weblog.html (accessed 20 Feb. 2004), n.p.

4. The term "intercreativity" was introduced by World Wide Web inventor Tim Berners-Lee in his book *Weaving the Web* (London: Orion Business Books, 1999).

5. John Hiler, "Borg Journalism: We Are the Blogs. Journalism Will Be Assimilated," *Microcontent News* (1 April 2002), http://www.microcontentnews.com/articles/borgjournalism.htm (accessed 27 Sep. 2004), n.p.

6. For more on the concept of the produser, in a journalistic context, see Axel Bruns's book *Gatewatching: Collaborative Online News Production* (New York: Peter Lang, 2005). Axel is currently extending his study of produsers and produsage in various contexts; more information can be found on his blog at http://snurb.info/. See especially "Some Exploratory Notes on Produsers and Produsage" (http://snurb.info/index.php?q=node/329), "Vlogging the Produser" (http://snurb.info/index.php?q=node/328), and the "Produsers and Produsage" section (http://snurb.info/index.php?q=node/453).

7. Clay Shirky, "RIP the Consumer, 1900–1999," *Clay Shirky's Writings about the Internet: Economics & Culture, Media & Community, Open Source*, May 2000, http://www.shirky.com/writings/consumer.html (accessed 31 May 2004), n.p.

SECTION ONE
BLOGS IN INDUSTRIES

The Practice of News Blogging

Axel Bruns

What we need ... is to revive notions of a republican community: a public realm in which a free people can reassemble, speak their minds, and then write or tape or otherwise record their extended conversation so that others out of sight might see it.

—James Carey[1]

As the broad range of perspectives assembled in this book demonstrates, the uses of blogs are many and varied. Among them, however, the practice of news blogging takes what is perhaps a somewhat special position—alongside only the traditional blog-as-diary approach, it is probably the most visible and best-known form of blogging.

> **News Blogging:**
> the practice of covering the news through blogging–whether by doing original reporting or by providing commentary on the news as it is reported in other news sources.

Put simply, news blogging is the practice of covering the news through blogging—whether by doing original reporting or by providing commentary on the news as it is reported in other news sources. This also explains why news blogging has become so well known in recent years: as blogging has established a more prominent place for itself in the public consciousness, mainstream journalists reporting on the phenomenon have naturally tended to look for prominent blogs in the topical area that was closest to their own interests—the news. Further, the continuing trend in journalism away from investigative reporting and toward pundit commentary also makes blog-based commentary on the news highly compatible with mainstream news content.

On the one hand, then, this positive bias among professional journalists to report about news-related uses of blogs might mean that the position of news-related blogging in relation to other blogging sub-genres has been somewhat overstated. Indeed, there are relatively few blogs that are exclusively devoted to news blogging, while by comparison there are many more blogs that are almost exclusively used as personal online diaries or for social interaction among friends. However, a great many blogs do cover the news at least from time to time, when it becomes relevant and important for their owners to do so. Blogger and journalist J.D. Lasica refers to such *ad hoc* news blogging as "random acts of journalism": he believes that "that's the real revolution here: In a world of micro-content delivered to niche audiences, more and more of the small tidbits of news that we encounter each day are being conveyed through personal media—chiefly Weblogs."[2]

This observation also points to the fact that news-related blogs are read in a mode that is very different from that commonly associated with the newspapers and bulletins of traditional journalism: few readers of such news blogs will follow a specific blog day by day, in the same way that they might have subscribed to a newspaper or watched the evening television news. Instead, they are more likely to discover news-related blog entries on individual bloggers' sites through other mechanisms: through links within the blogosphere itself, through meta-blogging tools such as *Technorati*, *Daypop*, or *Blogdex* (as well as news aggregators like *GoogleNews*), through scanning headlines on their favorite RSS feeds, or simply through information searches. Just as much as most news-related blog entries may constitute random acts of journalism, their discovery and reading tends to be just as random. We should note here that this does not in itself undermine the validity or importance of such news-related blog entries as a form of news reporting. However, as is frequently the case in the blogosphere, it is not the individual blog entry that is of foremost importance, but the (ideally interlinked) collection of blog entries on a shared topic, across the blogging world.

Practices in News Blogging

In their coverage of the news through their random acts of journalism, then, a variety of approaches can be observed. These range from first-hand reportage which is not unlike similar practices in professional journalism to a practice of what we will call gatewatching, which focuses more on providing additional commentary on the news as it is reported elsewhere.

If we conceptualize news as spanning a scale reaching from key world events through to the most mundane stories that affect only a handful of people, then first-hand reporting in news-related blogs can usually be found at either extreme of this scale. So, as Boczkowski notes, online news often "appears to present a micro-local focus, featuring content of interest to small communities of users defined either by common interests or geographic location or both."[3] Reporting such micro-news is a particular strength of blogs, as the bloggers doing so are often directly engaged in the news events themselves and therefore have access to more first-hand information as well as more interest in and knowledge of micro-news than the journalists employed by local news outlets.

Micro-News

Micro-news reporting through blogs can be regarded as a form of participatory journalism, then—a form of journalism where citizens participate as equal partners in the reporting of news. As Gans writes, this is a somewhat alien concept for traditional journalism, where often "journalists treat participators as deviants rather than as citizens, and whether they intend to or not, the news media discourage participation more than encourage it. Participatory news requires a reversal of these practices and should rest on the assumption that citizens are as relevant and important as public officials."[4]

Gans also points out that such *participatory journalism* should not be confused with the concept of *public journalism*, which has been trialed by some news organizations in an effort to arrest readership declines and create a more inclusive feel for news publications. As he describes it, "public journalism privileges mainstream issues, prefers mild controversies, and is unlikely to go beyond the ideological margins of conventional journalism. In contrast, I see participatory journalism as more citizen oriented, taking a political, and when necessary, adversarial, view of the citizen-official relationship."[5] In distinction from the public journalism model, then, the participatory journalism approach has more clearly activist elements: it "should also include news that is directly helpful in mobilizing citizens. ... If the strategies available to professional politicians are newsworthy, so are the strategies open to citizens."[6]

Unfolding World Events

By contrast, first-hand news reporting through blogs has also been prominent in the coverage of some of the key world events in recent times, from the September 11, 2001, attacks on the World Trade Center in New York to the subsequent wars in Afghanistan and Iraq, to the Christmas 2004 tsunami in the Indian Ocean and the London bombings in July 2005. As Mitchell describes it, "especially when big news breaks, it's tough to beat a Weblog"[7]— while the mainstream news channels and publishers are still in the process of scrambling their camera teams, re-arranging their Websites, or establishing lines of communication to their journalists in the field, in such major events bloggers are usually already on the scene and posting live updates to their sites.

In recent years, this trend has also been aided by the increasing availability of modern digital and wireless communications technologies. So, for example, the first photos from the aftermath of the London attacks were shot on mobile phone cameras and messaged directly to blogs and image sites such as *Flickr*, while workers in offices adjacent to the blast sites, or tourists in Internet cafés near the Thai beaches swamped by the tsunami waves, were able to post updates immediately as events unfolded. Such unedited, first-hand accounts have also come to have significance beyond reporting the news and contacting friends and family, in fact—so, for example, investigators into the London bombings used mobile phone footage by survivors to reconstruct the timeline of events, while the U.S. Library of Congress has compiled an archive of online reports following the September 11 attacks as an important historical record.

News Commentary

While it is true, then, that "with the ability to publish words and pictures even via their cell phone, citizens have the potential to observe and report more immediately than traditional media outlets do,"[8] such first-hand reporting practices nonetheless do not constitute the majority of news-related blogging, since they depend on the accident of the individual blogger's being at the scene of an event as it unfolds, or on the blogger's planning to attend and cover micro-news events. By contrast, a majority of bloggers still remain likely to encounter news events in a mediated form, through reports in the online and offline media (including other news-related blogs).

Such mediated access to the news does not rule out their own participation in the continued coverage of events and issues in the news, however—no more than it rules out commercial journalists' coverage of news as it is reported to them through the wire services. News blogging in this context—which makes up the majority of the news scale between micro-news and world events—engages predominantly in the commentary on, and collation and annotation of, news reports in other news sources: commonly, bloggers briefly summarize the issue or event in their blog posting (where possible linking to other blogs' or professional news sites' reports) before adding their own views or drawing connections between issues that appear to have been underrepresented in existing reports.

Writing in 2002, Hiler noted that there may also be a natural trajectory between these two forms of news blogging:

> I've met a good number of New York bloggers, and many of them have told me the same thing: "I had to start blogging after 9-11, just so my friends and family knew I was ok. Also, for the first time I felt like I had something to say—something worth blogging."
>
> There are so many post 9-11 weblogs that they've gotten their own name: *Warblogs*. Eight months after September 11th, Warblogs mostly dissect and analyze the news from the War on Terrorism. But immediately after 9-11, Warblogs provided a powerful form of personal journalism that captivated thousands of readers.[9]

This form of news blogging as commentary and annotation is especially powerful also in the coverage of continuing events, since the Weblog format's underlying model of posting *ad hoc* updates is better suited to continuous coverage. By contrast, print news must almost always publish whole newspapers for logistical reasons, while broadcast news must always repeat what it has already covered, since the ephemerality of the medium means that new viewers have no way of catching up with past coverage. News blogs, on the other hand, can publish even minor updates as soon as they come to hand, and serve as their own archives so that new readers can backtrack to the original report before reading more recent updates.

Blog-based news commentary may be a significant contributor to the overall rise of blogs as an informational medium, and its growing recognition by the wider public can again be linked to some key events—in the United States, for example, especially the 2004 presidential campaign. As a study by the Pew Internet & American Life Project reports,

> by the end of 2004 32 million Americans were blog readers. Much of the attention to blogs focused on those that covered the recent political campaign and the media. And at least some of the overall growth in blog readership is attributable to political blogs. Some 9% of internet users said they read political blogs "frequently" or "sometimes" during the campaign.[10]

Gatewatching

As blogger Glenn Reynolds describes it, then, although some bloggers "do actual reporting from time to time, most of what they bring to the table is opinion and analysis—punditry."[11] This focus on news commentary only mirrors a similar shift that has occurred in mainstream journalism for some time now—here, too, especially smaller operators have increasingly exchanged independent reporting for a greater reliance on sourcing content from the wire services (at best) and corporate and government PR departments (at worst). It would therefore be disingenuous to single out bloggers as regurgitators of second-hand news when in reality this practice is far more widespread. At the same time, some commentators have noted that in a time of immediate access to original information sources (especially through the World Wide Web), a fundamental change in the role of journalists is inevitable. For example, Bardoel and Deuze suggest that "with the explosive increase of information on a worldwide scale, the necessity of offering information about information has become a crucial addition to journalism's skills and tasks ... . This redefines the journalist's role as an annotational or orientational one, a shift from the watchdog to the 'guidedog.'"[12]

This fundamental challenge to traditional journalistic practices can be summarized as a shift from what has long been known as gatekeeping to a new practice of *gatewatching*. Instead of firmly keeping the gates—that is, making a selection of what news will or will not be seen by audiences—journalists as well as others engaged in reporting and discussing the news (such as bloggers) can now only *watch* the gates through which information passes from news sources

> **Gatewatching:**
> the observation of the output gates of news publications and other sources, in order to identify important material as it becomes available.

to the wider public, and can provide a suggestion of what they feel may be the most relevant news to their audience (knowing full well that many further news reports are also available to them). Instead of following the *New York Times* motto of providing "*all* the news that's fit to print" (which was always open to question, of course), it is now possible only to provide pointers to the news that may be *most important* to read, in the journalist's or blogger's judgment.[13]

This move to gatewatching over gatekeeping thus significantly reduces the power of the journalistic profession to affect public opinion. Gatekeeping in the mass media age exerted a measure of control over the public arena, but when gatekeepers lose their power to control the content of that symbolic arena, and when they are joined by an influx of alternative gatewatchers, "shared decision-making at the stage of gate-keeping changes the journalistic power balance ... and demonstrates a reconfigured world order at the press/media power centres, leading to a blurring of lines between the centre and the periphery in a critical journalistic practice."[14]

Further accelerating this shift, gatewatching is iterative: the material passing through the output gates of news blogs is further watched as potential source material by other gatewatchers. News bloggers therefore form a distributed community of commentators who will engage with one another's views on the news as much as with those expressed in other news sources. As Shirky describes it, "the order of things in broadcast is 'filter, then publish.' The order in communities is 'publish, then filter.' ... Writers submit their stories in advance, to be edited or rejected before the public ever sees them. Participants in a community, by contrast, say what they have to say, and the good is sorted from the mediocre after the fact."[15]

Weblog News as Multiperspectival News

While the point can be made that "the posting of established media source material does raise the question of whether this simply re-legitimises those media as the authentic forum for news,"[16] it is also important to note that the framing of that source material in a blog context creates a very different sense of the news than may exist in traditional journalistic publications. The focus on commentary and a kind of annotation at a distance of news reports found elsewhere on the Web, and the interlinkage and engagement between individual bloggers which is a fundamental feature of the blogosphere, turn Weblog news into a far more discursive form of news reporting than can usually be found elsewhere. In doing so it approaches what Heikkilä and Kunelius have described as dialogic or deliberative news: they postulate that

> [dialogic] journalism must openly encourage different readings (and search for new modes of stories that do so) and it must commit itself to [the] task of making these different readings and interpretations public. The challenge is to make the accents and articulations heard, to give them the power and position they need to argue on particular problems and to make them the objects and starting points for new emerging public situations and conversations.[17]

Deliberative journalism further advances beyond this by not only presenting these different readings, but enabling them to engage with one another directly and contribute to public deliberation: it "would underscore the variety of ways to frame an issue. It would assume that opinions—not to mention majorities and minorities—do not precede public deliberation, that thoughts and opinions do not precede their articulation in public, but

> **Deliberative Journalism:** journalism that enables a conversation between different viewpoints without privileging one as being more informed than another, and that aims to develop rather than merely express participants' opinions.

that they start to emerge when the frames are publicly shared."[18] This removes distinctions of status and expertise from the participants in the deliberation, much as news blogging undermines the privileged position of professional journalists as commentators on the news: "in a deliberative situation expert knowledge has no privileged position. All the participants are experts in the ways in which the common problem touches their everyday lives. Thus, opinions and knowledge expressed in deliberation articulate the experiences of the participants."[19]

In contrast to the conflict-based stories of the mainstream media, such deliberative news coverage begins to realize a form of news that journalism scholar Herbert Gans has envisaged for some three decades: multiperspectival news. As he describes the concept, "ideally, multiperspectival news en-

> **Multiperspectival News:** news that represents as many perspectives as is possible and feasible.

compasses fact and opinion reflecting all possible perspectives. In practice, it means making a place in the news for presently unrepresented viewpoints, unreported facts, and unrepresented, or rarely reported, parts of the population. To put it another way, multiperspectival news is the bottoms-up corrective for the mostly top-down perspectives of the news media."[20] It is not difficult to see news blogs as contributing to this goal.

News Blogs and the Mainstream News Media

Indeed, the rise of news blogging may be seen as a direct expression of news audiences' desire to increase the range of perspectives on the news that are available in the media. As a result, Bowman and Willis believe that "what is emerging is a new media ecosystem ..., where online communities discuss and extend the stories created by mainstream media. These communities also produce participatory journalism, grassroots reporting, annotative reporting, commentary and fact-checking, which the mainstream media feed upon, developing them as a pool of tips, sources and story ideas."[21] By now, even beyond the first-hand coverage of unfolding events which we have already discussed, several cases have indeed emerged where coverage of news through blogs has shown an effect on the mainstream media's news coverage as well.

News-related blogs and other alternative news Websites could therefore be seen as a second tier to the news media system, acting as a corrective and companion to the traditional news media. This model is remarkably similar to the two-tier system that Herbert Gans envisaged in 1980 in his discussion of multiperspectival news. He conceived of a media

> model, which combines some centralization and decentralization ... . Central (or first-tier) media would be complemented by a second tier of pre-existing and new national media, each reporting on news to specific, fairly homogeneous audiences. ... Their news organizations would have to be small [for reasons of cost]. They would devote themselves primarily to reanalyzing and reinterpreting news gathered by the central media—and the wire services—for their audiences, adding their own commentary and backing these up with as much original reporting, particularly to support bottom-up, representative, and service news, as would be financially feasible.[22]

In this environment, he writes, media in the second tier "would also function as monitors and critics of the central media, indicating where and how, by their standards, the central media have been insufficiently multiperspectival."[23] While clearly the technological and institutional setup of this second tier would have been impossible to foresee at the time, blogger commentary on the news appears to serve exactly these functions.

Today, as Lasica describes it, "on almost any major story, the Weblog community adds depth, analysis, alternative perspectives, foreign views, and occasionally first-person accounts that contravene reports in the mainstream press."[24] Journalists' response to such new competition and commentary from outside the profession has been mixed:

> in the United States news organizations responded to the challenge by questioning what non-news people have decided to call "news" ... and have noted that newspapers' versions of news is [sic] purer than the new versions because their news is edited and compiled according to various journalistic standards, such as impartiality. ... Such defensiveness reveals the extent to which online communication technologies—which

give all who own them the chance to be mass communicators—threaten traditional bastions of power.[25]

Indeed, the attacks by professional journalists (and some journalism scholars) against user-driven news reporting and commentary through blogs are often in direct parallel to the arguments (against all evidence) for why open source software development cannot possibly produce outcomes that are on par with commercial products. But what open source software development has already demonstrated is that Eric Raymond's theorem "given enough eyeballs, all bugs are shallow" tends to hold remarkably well. In other words, given a sufficient number of participants in a quality control process, significant errors will be identified and corrected. The theorem can be applied to news-based blogging just as well—and indeed the intercast of blogs in the blogosphere on a specific news topic constitutes exactly the form of fact-checking and exploration of backgrounds and motives that traditionalists claim is impossible without journalists.

What open source and news blogging do take for granted is that participants will exercise their own common sense in engaging in the software or news production process, however—and this seems to point to a significant distinction from professional journalism, which on average tends to have a relatively low regard for the intelligence of its audiences. Indeed, as Rushkoff puts it somewhat polemically, "the true promise of a network-enhanced democracy lies not in some form of web-driven political marketing survey, but in restoring and encouraging broader participation in some of the internet's more interactive forums. ... The best evidence we have that something truly new is going on is our mainstream media's inability to understand it."[26]

(In fairness, some of the more progressive news organizations have begun to see the benefits of catering for and engaging with news bloggers, though. For example, *The Guardian* newspaper and *BBC Online News* both run some blogs of their own and offer newsfeeds that bloggers can incorporate into their own sites, thereby making it easier for news bloggers to cite from and link to these sources.)

From Lecture to Conversation, from Users to Produsers

As blogger and journalist Dan Gillmor puts it, then, "if contemporary American journalism is a lecture, what it is evolving into is something that incorporates a conversation and seminar. This is about decentralization."[27] By making the news more accessible, discursive, and interactive (and for active bloggers even intercreative, since they can now contribute rather than merely receive news), news-related blogging might in fact generate more public interest in the news again, rather than making the public "tune out" as they are overwhelmed with a multiplicity of perspectives on the news.

Engagement with the news is, and should be, an inherently social activity— but for a long time now the necessities and limitations of print and broadcast media, and the standard institutional structures of mainstream news organizations, have restricted the social aspects of news engagement to the sidelines of the news process, focusing on information dissemination rather than on a public deliberation on news topics. The news engagement practiced in blogs and other collaborative news Websites, on the other hand, is closer to what Hartley describes as redactional journalism: for him, "reporting is the processing of existing discourse. But redactional journalism is not dedicated to the same ends as public-sphere journalism inherited from previous media; it doesn't have the same agenda-setting function for public affairs and decision-making as does traditional editing by editors."[28] Ultimately, then, if redactionary approaches multiply, it may be the case that "even as its representative democratic function is superseded, journalism itself massively expands."[29]

This suggests the possibility that we might "move away from the notion that journalism is a mysterious craft practiced by only a select priesthood—a black art inaccessible to the masses," as Lasica writes.[30] Regardless of whether or not the term "journalism" is applied to the practice of news-related blogging, a strict dichotomy between blogs and journalistic publications is no longer feasible in this redactionary environment. Surely not every blogger is a journalist, all of the time, but (also thinking back to Lasica's description of news-related blog entries as "random acts of journalism") many bloggers provide journalistic commentary on occasion—and this means that they step out of the news audience and into the community of news publishers.

Rushkoff similarly notes that "deconstruction of content, demystification of technology and finally do-it-yourself or participatory authorship are the three steps through which a programmed populace returns to autonomous thinking, action and collective self-determination."[31] In becoming active news publishers, commentators, and discussants, then, bloggers turn from users of the news to what we described in the introduction to this collection as "produsers"—a hybrid of producer and user.[32] If Rushkoff is right, their deliberative engagement as produsers of the news looks set to have effects well beyond the realm of news in its narrow definition, however. While news is the lifeblood of democracy, a side-effect of the strong dichotomies between producers and audiences for news has been to turn citizens into passive audiences for, rather than active participants in, democracy, able merely to switch channels every few years by voting in elections. If they are now becoming engaged produsers of the news, there is also a good chance for this change to affect their overall place in the democratic process. They could once again become active participants, users and producers or, in short, *produsers* of democracy as well.

NOTES

1. Cited in Jay Black (ed.), *Mixed News: The Public/Civic/Communitarian Journalism Debate* (Mahwah, N.J.: Lawrence Erlbaum, 1997), p. 14.

2. J.D. Lasica, "Blogs and Journalism Need Each Other," *Nieman Reports* (Fall 2003), http://www.nieman.harvard.edu/reports/03-3NRfall/V57N3.pdf (accessed 4 June 2004), p. 71.

3. Pablo J. Boczkowski, "Redefining the News Online," *Online Journalism Review*, http://ojr.org/ojr/workplace/1075928349.php (accessed 24 Feb. 2004), n.p.

4. Herbert J. Gans, *Democracy and the News* (New York: Oxford UP, 2003), p. 96.

5. Gans, *Democracy and the News*, pp. 98–99.

6. Gans, *Democracy and the News*, p. 96.

7. Bill Mitchell, "Weblogs: A Road Back to Basics," *Nieman Reports* (Fall 2003), p. 65.

8. Shane Bowman and Chris Willis, *We Media: How Audiences Are Shaping the Future of News and Information* (Reston, VA: The Media Center at the American Press Institute, 2003), http://www.hypergene.net/wemedia/download/we_media.pdf (accessed 21 May 2004), p. 47.

9. John Hiler, "Blogosphere: The Emerging Media Ecosystem: How Weblogs and Journalists Work Together to Report, Filter and Break the News," *Microcontent News: The Online Magazine for Weblogs, Webzines, and Personal Publishing*, 28 May 2002, http://www.microcontentnews.com/articles/blogosphere.htm (accessed 31 May 2004), n.p.

10. Lee Rainie, "The State of Blogging," *Pew Internet & American Life Project*, 2 Jan. 2005, http://www.pewinternet.org/pdfs/PIP_blogging_data.pdf (accessed 5 Jan. 2005), p. 1.

11. Qtd. in Bowman and Willis, *We Media*, p. 33.

12. Jo Bardoel and Mark Deuze, "'Network Journalism': Converging Competencies of Old and New Media Professionals," *Australian Journalism Review* 23.3 (Dec. 2001), p. 94.

13. For an extended discussion of gatewatching, see my book *Gatewatching: Collaborative Online News Production* (New York: Peter Lang, 2005).

14. Sujatha Sosale, "Envisioning a New World Order through Journalism: Lessons from Recent History," *Journalism* 4.3 (2003), pp. 386–87.

15. Clay Shirky, "Broadcast Institutions, Community Values," *Clay Shirky's Writings about the Internet: Economics & Culture, Media & Community, Open Source*, 9 Sep. 2002, http://www.shirky.com/writings/broadcast_and_community.html (accessed 31 May 2004), n.p.

16. Graham Meikle, *Future Active: Media Activism and the Internet* (New York: Routledge, 2002), p. 100.

17. Heikki Heikkilä and Risto Kunelius, "Access, Dialogue, Deliberation: Experimenting with Three Concepts of Journalism Criticism," *The International Media and Democracy Project*, 17 July 2002, http://www.imdp.org/artman/publish/article_27.shtml (accessed 20 Feb. 2004), n.p.

18. Heikkilä and Kunelius, n.p.

19. Heikkilä and Kunelius, n.p.

20. Gans, *Democracy and the News*, p. 103.

21. Bowman and Willis, *We Media*, p. 13.

22. Herbert J. Gans, *Deciding What's News: A Study of* CBS Evening News, NBC Nightly News, Newsweek, *and* Time (New York: Vintage, 1980), p. 318.

23. Gans, *Deciding What's News*, p. 322.

24. Lasica, "Blogs and Journalism Need Each Other," p. 73.

25. Mark D. Alleyne, *News Revolution: Political and Economic Decisions about Global Information* (Houndmills, UK: Macmillan, 1997), p. 33.

26. Douglas Rushkoff, *Open Source Democracy: How Online Communication Is Changing Offline Politics* (London: Demos, 2003), http://www.demos.co.uk/opensourcedemocracy_pdf_media_public.aspx (accessed 22 April 2004), pp. 53–54.

27. Dan Gillmor, "Moving toward Participatory Journalism," *Nieman Reports* (Fall 2003), p. 79.

28. John Hartley, "Communicative Democracy in a Redactional Society: The Future of Journalism Studies," *Journalism* 1.1 (2000), p. 44.

29. Hartley, "Communicative Democracy," p. 44.

30. Lasica, "Blogs and Journalism Need Each Other," p. 73.

31. Rushkoff, *Open Source Democracy*, p. 24.

32. For more on the concept of the produser in a journalistic context, see my book *Gatewatching: Collaborative Online News Production* (New York: Peter Lang, 2005). I am currently extending my study of produsers and produsage in various contexts; more information can be found on my blog at http://snurb.info/. See especially "'Anyone Can Edit': Understanding the Produser" (http://snurb.info/index.php?q=node/286), "Vlogging the Produser" (http://snurb.info/index.php?q=node/328), and the "Produsers and Produsage" section (http://snurb.info/index.php?q=node/453).

Journalists and News Bloggers: Complements, Contradictions, and Challenges

Jane B. Singer

L et's jump right to the subject of considerable debate: Are bloggers journalists?

Journalists' immediate answer is: No way. Bloggers' immediate answer is: No thanks. But the question demands more than a two-word response.
Just what, if anything, is it that journalists do that is different from what bloggers do? Are bloggers acting as journalists some of the time but as something
else (like what?) the rest of the time? And why should the rest of us care?

For starters, the majority of the millions of blogs out there today are not
journalistic because they provide no public service. The purpose of journalism,
at least in a democracy, is to serve the public by providing citizens with the information they need to be free and self-governing.[1] Most bloggers are folks for
whom the format is a fun, easy way to create a personal journal for their own
enjoyment or that of their family and friends. Enhancing democracy is not
high on the agenda.

But some bloggers clearly have something grander in mind, and they are
the ones we are concerned with here. These are the people who post, ardently
and often, about politics, government, war, the media, social issues, and a host
of other topics that also fill news holes in traditional media around the world.
And they have a following that extends well beyond their circle of personal ac-
quaintances. People turn to them for information, commentary, and insights—
the same sorts of things journalists provide.

My own sense, and that of other observers whom we'll get to soon, is that
these "news bloggers" and journalists fill sometimes-overlapping but essentially
different niches in the information environment. Their relationship is both
symbiotic and complementary: bloggers and journalists often irk one another,

but ultimately offer mutual benefits, and a twenty-first century public is better served by both together than by either alone.

This chapter starts with one fundamental way in which journalists differ from news bloggers: their nearly opposite approaches to truth. A look at the synergies between bloggers and journalists follows. With most of the chapter exploring the two as distinct, if complementary, types of communicators, "j-bloggers," or journalists who blog, merit some attention at the end.

Seeking and Reporting Truth

Early in 2005—barely two months after a U.S. presidential campaign in which news bloggers propelled themselves into the national consciousness through nonstop flogging of both candidates and the media, and just weeks after bloggers provided more vivid coverage of the South Asian tsunami disaster than any but the largest media outlets could begin to approach—media scholar Jay Rosen declared that "journalists vs. bloggers is over." Actually, he mostly said that journalism itself is over, at least as it has been traditionally constructed. What many in the media affectionately (or not) call "Big-J Journalism" has lost great people, who have deserted "when they ran out of room for their ideas." It has lost credibility, with the notion of objectivity under fire and large numbers of people hooting in derision at the idea of a neutral, nonpartisan press. In a world in which anyone can publish anything, and without any gates for information gatekeepers to guard, it has lost its franchise.[2]

The points are well taken, but rather than signaling an imminent obituary, they encourage a closer look at what journalism really is or does. A great many definitions have been offered. Journalists see journalism as, among other things, a mirror, a container for the day's events, and a story about those events; scholars refer to a profession, an institution, a text, a practice.[3] To me, journalism is a particular approach to obtaining and communicating civically important truth. It is not the only approach by any means, and in our open and exponentially extended information environment, it may even seem like an outmoded one—not unreasonably, since it dates to the eighteenth century.

Journalists' route to truth stems from a set of Enlightenment ideas, including that people are inherently rational, and that reality is both observable and verifiable. Journalists understand truth as something that can be seen or heard, by themselves or by someone else who is a reliable source, and that can be verified and corroborated. A great many factors play into the news selection process, but at a basic level, what journalists do is collect and vet information as best they can before deciding whether to pass it along to the public.[4] They then offer as truth information that survives what ideally is rigorous scrutiny by both reporter and editor, scrutiny in which personal beliefs are set aside in favor of nonpartisan fairness. The essence of journalism is this process of verification.[5]

The fact that other information is also available does not change this basic definition of journalism. Other information has always been available; it's just that there is now much, much more of it, and it is much, much easier for anyone to generate and disseminate. Arguably, this makes journalism more valuable, not less so. Journalistic gatekeeping is no longer a matter of determining which items are to be allowed to circulate; it is a matter of certifying that among the millions of freely circulating items, some subset has been independently verified as trustworthy.

This is not to say that the process works flawlessly or that even the journalists who diligently follow it (let alone the ones who don't) necessarily generate truth. Obviously, neither is the case. Nor is it to say that this is the only route to truth. But as a way to define the role and process of journalism in a democracy, it seems to serve pretty well.

Bloggers also value truth, but they have quite a different view of how to get there. Theirs is a more postmodern approach, acknowledging that everyone holds his or her own version of the truth; brought together, those views form a subjective, multi-faceted but cohesive whole. Bloggers place a premium on the power of the collective, of shared knowledge and the connections among those who possess and are willing to exchange it.

The blogosphere is an electronically enabled marketplace of ideas. In it, the vetting process is social and public, not individual and relatively private as it is for journalists—and the vetting takes place after publication, not before. The blog is an open forum in which information is offered, revised, extended, or refuted; the more who participate in the process of generating truth, the merrier. It's John Milton with moxie and a modem: get all the ideas on the table, kick them around, and individual falsehood gives way to collectively derived truth. The whole knows more, and better, than any of its component parts. Other participatory media forms, such as wikis, operate on much the same principle.

Of course this process doesn't work flawlessly, either. The collective is neither necessarily wiser than the individual, nor the majority more cogent than the minority. Even if truth eventually does emerge from the conversation, a whole lot of potentially harmful untruths can be published, globally, during the kicking-around phase. But the priority given to interactive conversation, rather than the one-way lecture delivered through the newspaper or television newscast,[6] highlights this view of truth and how to attain it.

Symbiotic and Complementary

Which way is better? That's a question destined to lead to little except name-calling; both journalists and news bloggers can and do adamantly claim superiority by pointing out the strengths of their approach and the faults of the other. Journalists who spend their evenings complaining to one another over beers about how their editor butchered their story write ardent defenses of the editing process itself as the guardian of information integrity. Bloggers who spend their evenings—as well as their mornings, afternoons, and nights—talking incessantly among themselves appear on talk shows to laud the value of a place where everyone's voice is heard.

In fact, both journalists and news bloggers are and will remain valuable, to one another and to the public. They offer, as I suggested at the start, symbiotic and complementary approaches to information.

That news bloggers need journalists is obvious—where else would they find new things to talk about if not in the "MSM," their often-derogatory term for the mainstream media? Although there are a growing number of exceptions, relatively few bloggers actively gather original information themselves. The good ones monitor a range of media outlets; compare their versions of events, and compare present versions with previous ones; root around online for alternative perspectives; then post what they have learned, link to what they have found, and offer their opinions about it all. The less good ones jump right to the part where they offer their opinions.

News bloggers also need journalists to give them publicity. Who, besides other bloggers and perhaps a stung politician or two, would even know what the fuss was about if not for widespread media coverage? Journalists write about news blogs, quote from them, compete with them, follow their leads—and create their own blogs as well. (More on that last one in a bit.)

And news bloggers need journalists, frankly, as a foil. Public displeasure with the MSM is no secret. Surveys indicate that majorities of Americans believe the news media are biased, make up stories,[7] and cannot be trusted to tell the truth[8]; more than a third see them as outright immoral.[9] News bloggers—many of whom seem to share these attitudes passionately—point to their own practices as an alternative or even an antidote; indeed, they are quite articulate about expressing their belief that their approach to truth, as outlined above, is superior.

That said, news bloggers have much to learn from journalists—including some of their much-maligned practices and much-derided ethical guidelines. The checks and balances that (yes) an editor provides, the enhanced credibility that comes from doing one's own reportorial legwork, the development of a cogent writing style, a working knowledge of valuable tools such as the U.S. Freedom of Information Act, a commitment to accuracy, and an avoidance of potential conflicts of interest—all are journalistic lessons that news bloggers can profitably take to heart. "Journalists, as members of the 'Fourth Estate,' have long held power. Now bloggers are positioned to share some of that. Take care, please," urges longtime print and online journalist Steve Outing,[10] who also is a senior editor for the Poynter Institute.

Actually, a number of folks have taken a cut at drafting guidelines for news bloggers—and they sound a whole lot like guidelines for journalists. For instance, those offered by Jonathan Dube on the *Cyberjournalist* site draw heavily on the Society of Professional Journalists code of ethics; he urges bloggers to minimize harm, to be accountable, and to be honest and fair.[11] Blogging pioneer Rebecca Blood, in her *Weblog Handbook*, proposes a set of "Weblog ethics" that emphasize transparency, including disclosure of any conflict of interest and of any "questionable and biased sources," and exhort bloggers to "publish as fact only that which you believe to be true."[12] *Online Journalism Review* suggests telling readers how information was obtained, clarifying commercial relationships, staying away from plagiarism and conflicts of interest, and checking out information before publishing it.[13] And so on.

Besides, it's only a matter of time—which may already have arrived as you are reading this—before a blogger is rudely awakened by judge or jury to some facts of life that journalists have drilled into them from the first day of j-school, facts such as what constitutes libel or plagiarism or invasion of privacy. One more thing bloggers can learn from journalists is the name of a good libel lawyer.

The ways in which journalists need news bloggers may be subtler; after all, journalists have managed without this particular swarm of gadflies for hundreds of years. But viewing these bloggers as pests misses the real values they provide.

For starters, bloggers' emphasis on what they call transparency, somewhat akin to the journalistic notion of accountability, is worth paying attention to. Journalists have become far too willing to use anonymous sources, particularly in their coverage of politics, policy, and government. Many appear trapped by professional norms of objectivity that too easily devolve into "he said, she said" reporting,[14] leaving them vulnerable to manipulation by those sources and abandoning the public a long way from the truth that the norms are supposed to protect. And they seem to have found no better way to convince people of their commitment to fairness and nonpartisanship than to earnestly plead, "trust us, we really do set aside our personal feelings when we do our jobs."

Most probably do, but as we've seen, a precipitously plummeting number of people believe it. The full disclosure philosophy of bloggers is not perfectly suited to journalists, but a little more openness would go a long way. News bloggers are transparent not only in their motive but also in their process, extensively using links to documents, sources, news articles, and other sorts of evidence to buttress their points and establish their authority.[15] Journalists would do well to take the cue, especially in the online presentations of news but also in the types of support they offer for stories in other media. In doing so, they would greatly expand the number and diversity of their sources, another bonus of news blogging.

Journalists also can benefit from the fact that news bloggers offer thousands of extra pairs of ears to the ground. Bloggers notice, and write about, things that journalists may miss, and some of those things are bound to be important. One of the most famous examples is bloggers' role in the political demise of U.S. Senate Majority Leader Trent Lott, who a few years ago told a public gathering that the country would have been better off had voters had the wisdom to elect segregationist Strom Thurmond as president back in 1948. Bloggers weren't there—but they were watching the C-Span cable network, which was. Several A-list news bloggers kept Lott's comments front and center on their blogs until mainstream journalists noticed; within weeks, Lott was the former Senate Majority Leader.[16] In the years since, there have been numerous other examples of news bloggers flogging a story until it makes the MSM agenda.

In fact, a fair number of those stories are about the media, a favorite subject of bloggers everywhere—which brings me to a major benefit that news bloggers offer journalists, though it may sometimes seem more a curse than a blessing. Bloggers have gleefully taken on the role of watchdog on the watchdog. Some very big dogs have felt the bloggers' bite, from (former) CBS News anchor Dan Rather, who took the fall for mishandling of a story about President George W. Bush's disputed Air National Guard service as a young man, to (former) *New York Times* Executive Editor Howell Raines, turned to toast in the blogosphere for his role in the Jayson Blair scandal.

Bloggers may not individually have the power of a Rather or a Raines, but they do have influence—the ability, as *InstaPundit* news blogger Glenn Reynolds puts it, to "get ideas noticed that would otherwise be ignored and to shame people into doing their jobs better."[17] People such as journalists. Blogs can serve as a corrective mechanism for sloppy or erroneous reporting.[18] "For lazy columnists and defensive gatekeepers, it can seem as if the hounds from a mediocre hell have been unleashed," *Columbia Journalism Review* contributor Matt Welch writes[19]; journalists can now expect that someone out there is going to fact-check just about anything they write. Such ardently attentive scrutiny can seem brutal, but it has the wonderful benefit of making an editorial stance of arrogance or aloofness hard to sustain.[20] Call the news blogosphere "Estate 4.5," neither part of government nor of the media but keeping a wary and watchful eye on both, providing a valuable check against inadequacy and abuse of power by either.[21]

So news blogging and journalism are symbiotic forms, each needing and potentially benefiting from the other's presence and practice. They are also complementary. There is, in fact, no single route to truth; there may or may not be one truth, but it doesn't require a metaphysical leap to recognize that there are many ways to attain it (or them). News bloggers and journalists fill different niches in the information space. Bloggers can focus narrowly on stories that may fly under the radar screen of Big-J Journalism and can doggedly pursue an agenda that they find personally important—and that, like the story of Lott, sometimes turns out to be socially important as well. Journalists can bring the power of their institution, reputations, and professional standards to bear on stories of broad public interest. Together, they can do a more thorough job of serving the public, with the strengths of each correcting the shortcomings of the other.

Journalism vs. blogging is, indeed, over—not with victory but with, ideally, peaceful coexistence. In his book *Mediamorphosis*, Roger Fidler points out that rarely do media forms supplant one another.[22] Older forms typically do not die out, but they do evolve. Challenged by a successful newcomer, the people who love them and are adept at working within them do a little soul-searching to figure out what it is that they really do best. Then they decide how to go about doing it in the changed media environment that the new form has brought into being.

Perhaps the greatest benefit of all those that news bloggers bring to journalists is the impetus to conduct this soul-searching, to reaffirm their commitment to the public good, to recognize the strengths of their approach to meeting that commitment—and to go out and do it better.

Postscript: Journalists Are Bloggers, Too

The picture of two distinct groups of people, the bloggers and the journalists, is actually a little fuzzier than the previous discussion might suggest. So I'll end with a few words about journalists who blog.

Actually, even many folks who consider themselves full-fledged bloggers are being treated, in practical terms, as journalists. For instance, they are getting press credentials to cover major events, perhaps most prominently the 2004 U.S. political conventions. Some news bloggers are attracting advertisers—and struggling with commercial pressures to keep those advertisers happy. As of this writing, there are moves in the U.S. Congress to establish a national shield law safeguarding a journalist's ability to protect the identity of confidential sources. One version of the proposed law extends the definition of a news medium to "any printed, photographic, mechanical, or electronic means of disseminating news or information to the public," a construction that covers blogs and other Web-only news sites.[23]

In the meantime, a great number of journalists already are operating blogs—hundreds, according to the *Cyberjournalist* site. Some are blogging on their own, with or without the sanction of their employers. But most are "j-bloggers," typically columnists or top-level reporters whose byline gives them sufficient buzz to start their own blogs under the auspices of their media organization.

Blogs enable journalists to cover big stories in novel ways. Many journalists covering the 2004 U.S. political campaigns for major media outlets did so as both reporters and bloggers. *Newsweek* magazine blogged its coverage of celebrity decorator Martha Stewart's trial; *The New York Times* took on climate shifts in the Arctic. Pick a topic that a news organization is willing to devote resources to telling well online, and the odds are rapidly increasing that part of its coverage will include a blog. Editorial boards are also getting in on the act, using blogs as a way of showing readers how they make their decisions.[24]

The sailing has not been uniformly smooth. A number of reporters have found themselves in hot water for expressing opinions on their blog, thus blurring their profession's beleaguered but still entrenched premise of objectivity. CNN editors told a correspondent to stop blogging about his experiences covering the war in Iraq.[25] A *St. Louis Post-Dispatch* reporter resigned after coming under fire for writing a blog in which he lambasted the paper.[26] A *Houston Chronicle* bureau chief was fired after using his blog to criticize politicians he covered for the paper; what he saw as a harmless creative outlet struck his editors as an appalling conflict of interest.[27]

Despite the bumps, media organizations seem to be finding that news blogs are an intriguing way to engage readers and to cover their communities creatively. They allow journalists to share information that doesn't fit in the traditional format's limited news hole, to squeeze more voices into their reporting, and to get potentially valuable feedback from the public.

Just what the bloggers have been saying all along.

Are journalists bloggers? Growing numbers are, at least some of the time. Are news bloggers journalists? Not yet—and ideally, things will stay that way. We need the two sets of independent—in the best senses of the word—voices offering us alternative, complementary paths to truth.

NOTES

1. Bill Kovach and Tom Rosenstiel, *The Elements of Journalism: What Newspeople Should Know and the Public Should Expect* (New York: Crown, 2001), p. 17.

2. Jay Rosen, "Bloggers vs. Journalists Is Over," *PressThink* (15 Jan. 2005), http://journalism.nyu.edu/pubzone/weblogs/pressthink/2005/01/15/berk_pprd.html (accessed 10 Sep. 2005).

3. Barbie Zelizer, "Definitions of Journalism," *The Press*, Geneva Overholser and Kathleen Hall Jamieson, eds. (New York: Oxford UP, 2005), pp. 66–80.

4. Jane B. Singer, "The Marketplace of Ideas—with a Vengeance," *Media Ethics* 16.1 (2005), pp. 1, 14–16.

5. Kovach and Rosenstiel, *Elements of Journalism*, p. 71.

6. Dan Gillmor, *We the Media: Grassroots Journalism by the People, for the People* (Sebastopol, CA: O'Reilly, 2004).

7. Rachel Smolkin, "A Source of Encouragement," *American Journalism Review* (Aug./Sep. 2005), http://ajr.org/Article.asp?id=3909 (accessed 10 Sep. 2005).

8. Humphrey Taylor, "Trust in Priests and Clergy Falls 26 Points in Twelve Months," *The Harris Poll: Harris Interactive* (27 Nov. 2002), http://www.harrisinteractive.com/harris_poll/index.asp?PID=342 (accessed 10 Sep. 2005).

9. Project for Excellence in Journalism, "Overview: Public Attitudes," *The State of the News Media 2005: An Annual Report on American Journalism* (2005), http://www.stateofthemedia.org/2005/narrative_overview_publicattitudes.asp?cat=7&media=1 (accessed 10 Sep. 2005).

10. Steve Outing, "What Bloggers Can Learn from Journalists," *Poynteronline.com* (22 Dec. 2004), http://poynter.org/content/content_view.asp?id=75665 (accessed 10 Sep. 2005).

11. Jonathan Dube, "A Blogger's Code of Ethics," *Cyberjournalist.net* (15 April 2003), http://www.cyberjournalist.net/news/000215.php (accessed 10 Sep. 2005).

12. Rebecca Blood, *The Weblog Handbook: Practical Advice on Creating and Maintaining Your Blog* (New York: Perseus Books Group, 2002), http://www.rebeccablood.net/handbook/excerpts/weblog_ethics.html (accessed 10 Sep. 2005).

13. *Online Journalism Review*, "Ethics" (9 March 2005), http://www.ojr.org/ojr/wiki/ethics (accessed 10 Sep. 2005).

14. Brent Cunningham, "Re-thinking Objectivity," *Columbia Journalism Review* (July/Aug. 2003), http://cjr.org/issues/2003/4/objective-cunningham.asp (accessed 10 Sep. 2005).

15. J.D. Lasica, "Transparency Begets Trust in the Ever-Expanding Blogosphere," *Online Journalism Review* (12 Aug. 2004), http://www.ojr.org/ojr/technology/1092267863.php (accessed 10 Sep. 2005).

16. Tom Regan, "Weblogs Threaten and Inform Traditional Journalism," *Nieman Reports* (Fall 2003), pp. 68–70.

17. Rachel Smolkin, "The Expanding Blogosphere," *American Journalism Review* (June/July 2004), http://ajr.org/article.asp?id=3682 (accessed 10 Sep. 2005).

18. Paul Andrews, "Is Blogging Journalism?" *Nieman Reports* (Fall 2003), pp. 63–64.

19. Matt Welch, "The New Amateur Journalists Weigh In," *Columbia Journalism Review* (Sep./Oct. 2003), http://www.cjr.org/issues/2003/5/blog-welch.asp (accessed 10 Sep. 2005).

20. Bill Mitchell and Bob Steele, "Earn Your Own Trust, Roll Your Own Ethics: Transparency and Beyond," paper presented to the *Blogging, Journalism and Credibility* conference at Harvard University, Cambridge, MA (17 Jan. 2005), http://cyber.law.harvard.edu/webcred (accessed 10 Sep. 2005).

21. K. Daniel Glover, "Journalists vs. Bloggers," *Beltway Blogroll* (8 July 2005), http://beltwayblogroll.nationaljournal.com/archives/2005/07/journalists_vs.html (accessed 10 Sep. 2005).

22. Roger Fidler, *Mediamorphosis: Understanding New Media* (Thousand Oaks, CA: Pine Forge Press, 1997).

23. Reporters Committee for Freedom of the Press, "Federal Shield Law Efforts" (2005), http://www.rcfp.org/shields_and_subpoenas.html#shield (accessed 10 Sep. 2005).

24. Mark Glaser, "Papers' Online Units Allow Editorial Boards to Lift Veil with Video, Blogs," *Online Journalism Review* (9 March 2004), http://www.ojr.org/ojr/glaser/1078877295.php (accessed 10 Sep. 2005).

25. Barb Palser, "Free to Blog?" *American Journalism Review* (June 2003), http://ajr.org/article.asp?id=3023 (accessed 10 Sep. 2005).

26. David Kesmodel, "Should Newspapers Sponsor Blogs Written by Reporters?" *The Wall Street Journal* (12 July 2005), p. B1.

27. Steve Olafson, "A Reporter Is Fired for Writing a Weblog," *Nieman Reports* (Fall 2003), pp. 91–92.

CHAPTER FOUR

Publishing and Blogs

Joanne Jacobs

Become an expert in your field and book sales will follow... By projecting yourself as an expert in the genre in which you write, you can open new doors for networking, doors that often remain shut without that expertise status.

—Brett Sampson, Outskirt Press Book Publishing (2005)

sk an editor of a traditional book publishing house what the most difficult aspect of dealing with authors is, and they are likely to tell you that it's getting the authors to deliver their manuscripts on time. Whether it be procrastination or perfectionism that consistently delays deadlines, authors have a reputation in book

> **Publishing Blogging** (as opposed to **blog publishing**): blogging about the industry of traditional book publishing, or blogging that actually facilitates production of book content.

publishing for being decidedly unreliable in their capacity to deliver content. And even when they do deliver, the quality of the work submitted may be far from the expectations of the publishing house, whereupon editors are required to step in and act as a kind of diplomat between the warring parties of Author and Publisher. It is a delicate and politically fraught role, necessitating a balance between supporting the egotism of the author and honoring the commercial imperatives of the publishing house. And mediation of the issues between parties must all be conducted as quickly and efficiently as possible, ensuring that neither the author nor the publishing house miss their opportunity for competitive advantage. Failing to release a co-production before a competitor publishes something in a similar field can adversely affect both sales and reputation.

There is no doubt that the role of a traditional book publisher is particularly difficult. And where multiple authors are concerned—as in this book—the role becomes even more complex in ensuring a consistency of style and presentation throughout the title, and coordinating the efforts of many authors writing on significantly differing topics. So it is perhaps unsurprising that editors are beginning to use whatever tools they have at their disposal to ensure that their tasks are performed as efficiently as possible. Blogs are the latest in a long line of tools for editors in reducing the time-to-market costs of traditional book production. But they carry with them a stigma that has never before been associated with desktop publishing and other electronic tools. Where desktop publishing programs assisted formatting, design, and editing of content, and extranets, electronic mail, the World Wide Web, and other applications of the Internet assisted in the distribution of content, blogs actually challenge the dominant paradigm of technology as an agent of content professionalization. Blogs are generally perceived to invoke a culture of mass amateurization of content, where the value and credibility of the authorial voice is made vulnerable through commentary systems. Blog commentator Clay Shirky has argued that the lack of a publishing filter prevents the content made available through Weblogs from being regarded with the same *imprimatur* as traditionally published works. And while he argues that the collaborative aspects of

> **Intrinsic /
> Extrinsic Value:**
>
> intrinsic value is the social capital invested in an entity, as opposed to its published value or price (extrinsic value).

blog publishing have a profound effect on writing output and productivity, he also warns that the norm of the Web is based on mass amateurization; thus the efficiencies to be derived from the use of blogs in publishing can adversely affect the perceived value of the final work. Shirky argues that the value created by traditional publishing is partly intrinsic.

> It takes real work to publish anything in print, and more work to store, ship, and sell it. Because the up-front costs are large, and because each additional copy generates some additional cost, the number of potential publishers is limited to organizations prepared to support these costs. (These are barriers to entry.) And since it's most efficient to distribute those costs over the widest possible audience, big publishers will outperform little ones. (These are economies of scale.) The cost of print insures that there will be a small number of publishers, and of those, the big ones will have a disproportionately large market share.
> Weblogs destroy this intrinsic value, because they are a platform for the unlimited reproduction and distribution of the written word, for a low and fixed cost. No barriers to entry, no economies of scale, no limits on supply.[1]

The ease of blogs' publishing information online to a virtually limitless audience has brought about an environment where authors who may never have been published in traditional print are able to access audiences and generate a readership, almost regardless of the quality and credibility of their writing. Thus, of all industries, the traditional book, magazine, and cross-media publishing market is most likely to under-value the influence of blogs, and to depict the writing in blogs as less finished, less authoritative, and less considered than that which might be found in traditional works. But in practice, the opposite is true.

In the serious industry of book publishing, blogs are being integrated into all facets of the publishing process. From historical registers of industry news to ongoing discussions about publishing practices, and even to the process of content development, blogs are making their presence felt in the sector perhaps most challenged by the phenomenon. And the role of blogs in publishing is not merely a matter of experimentation. When used as an editing tool for book publishing, blogs are beginning to be implemented as a serious commercial strategy to reduce the risk associated with authors who fail to produce timely and adequate quality manuscripts for their publishing houses. As a tool, blogs don't act to formalize content delivery so much as to initiate and shape it.[2] Instead of making the text look more readable (as in the case of publishing applications), or permitting the submission of manuscripts in a manner that can be easily accessed and edited, blogs are being implemented much earlier in the development of content, so that editors and authors can ensure that the style, accuracy, and completeness of a work match the quality aspirations of publishing houses.

Blog Practices in Publishing

The range of uses of blogs in traditional publishing is quite diverse. Beyond the use of an extranet-hosted blog as a closed forum for content development among authors, reviewers, and editors, there are several other applications of the blog that are observable. In the first place, blogs about the practice of traditional publishing have emerged as a kind of knowledge management tool, where active publishers and editors report on publishing-related news and discuss issues facing the publishing industry, particularly from an economic or political perspective. Then there are blogs associated with the minutiae of editing and sharing details of updates in style guides and formatting. There are collaborative publishing blogs for works being developed by multiple authors. (These are covered in further detail below.) Finally, there is a growing trend in the use of blogs as a dynamic marketing tool to promote newly published works. These last blogs are designed to engage a readership in a manner that represents a high risk for publishers, as comments and reader-oriented blog posts can either positively or negatively affect reputation of the sister text. In the case of Howard Rheingold's *Smart Mobs* site,[3] the blog contributed toward the final copy of the 2002 book, but also expanded on the theories propounded in the text, and continues to be regarded by its readership as a growing knowledge and critical discussion base, where posts may well be adapted or copied whole into future editions of the printed edition. Indeed, the readers are always aware that their virtual presence through comments and participation in ideas generated through the *Smart*

Mobs site could be immortalized in future editions of the book or in alternative works by Howard Rheingold. This is the text-based equivalent of people standing in the background of an outside television broadcast, waving to the cameras. Blog readers of *Smart Mobs* tend to carefully post suggestions and content in the hope that their analysis may contribute to the future canon of the printed edition, thus obtaining a kind of added credibility to their words.

For each of these models, blogs represent a value-addition to standard industry reporting practices and processes. As *Online USA* editor Marydee Ojala has noted,

> blogging certainly has the potential to transform organizations by greatly accelerating the rate of information and knowledge exchange, allowing tacit knowledge to flow quickly to those who need it, when they need it.[4]

It is the flow and timeliness of blogs that provides for the traditional publishing sector the opportunity to improve on publishing output. And for each application of blogging, rather than devaluing the printed works being produced by traditional publishing houses, there is strong evidence to suggest that the marketing and demand for printed works grows with discussion of works online. And even if titles are not profiled, book publishing houses are using information gleaned from industry-related blogs to keep track of new releases, audience discussions, and demand for various genres, in order to monitor their place in the industry and maintain a sustainable competitive advantage.

Blogging the Industry

Where blogs are regarded to have had an early influence in traditional publishing is in the discussion and posting of news about the sector. At the time of writing, the search engine *Google* registered more than a thousand blogs reporting news about the book publishing industry, and that number grows daily. Regardless of any consideration of blogs as highly *ad hoc* and not as efficient as mainstream publishing tools, blogs have been readily adopted by mainstream publishing representatives as an inherently useful tool for identifying and disseminating knowledge about the industry. Technologically savvy publishing representatives are using blogs as a means of generating industry support for changes in editing styles and processes, as well as encouraging productivity and sales of traditional book publishing by generating interest in forthcoming publications.

And unlike USENET newsgroups and online forums, the structure of the blog makes this process of interest generation simpler. Because blog content is developed by just one or a few main bloggers, they are less exposed to the characteristic breakdown of groups that commentator Clay Shirky and others have found in their research on the politics of groups.[5] Instead of developing as dysfunctional cybercommunities of a few dominant voices, drowning out the issues of interest to an otherwise dedicated sector of readers, the act of maintaining control at least over the news published in industry blogs enables development of a sustained readership. As the collaborative writing process of a few key bloggers, blogs created by traditional publishing firms such as *thebookstandard.com* and *readersread.com* have managed to attract strong audiences and almost incidentally have created an historical register of happenings in the sector.

The incidental value of blogs in recording an industry history of events and happenings is another example of the kind of public benefits to be derived from the activity of users, as identified in the work of Hewlett-Packard researchers Scott Golder and Bernardo Huberman.[6] In their work on the collaborative tagging of content online, they note that users who bookmark and tag works found online do so primarily for their own benefit, and not for the collective good, but that the process of tagging such works nevertheless can constitute a useful public good. The same incidental public good can be traced to the activity of publishing industry blogging: the function of blogs as a dynamic annotated index of news for the industry acts itself as a published resource and reference for stakeholders in the sector. Where multiple authors may contribute to such an industry blog as occurs with *thebookstandard.com*, the collaborative publishing structure provides a model for new publishing processes.

Students of information economics will be familiar with the theories of network effects (sometimes called network externalities), whereby the value of information held by a single individual is increased by its being shared among others.[7] Information value-addition produces knowledge, and the value of knowledge increases as it is shared. But central to this notion is the sense that the value of knowledge can only be

> **Network Effects / Network Externalities:**
> a situation where the benefits of adopting a process or product are an increasing function of the number of other users of that process or product.

sustained if information is updated, renewed, and appropriately cross-referenced. In terms of collaborative publishing, this would tend to imply that the value of the content being produced by group publishing efforts will only be sustained if the productivity of group members is high and if there is a constantly renewable source of producers. This is where blog tools can act effectively between multiple authors of a traditional published text.

Blogediting

A growing application of the use of blogs in the traditional publishing sector is in the process of collaborative publishing, or "blogediting." In the development of non-fiction works in particular, the use of blogs as a drafting tool either in a closed forum with co-authors and copy editors, or even in an open forum, is a useful means of keeping track of the productivity of authors as well as a practical risk management tool, encouraging readers to link to appropriate resources to support statements made in chapters. It's this hypertextuality and link farming in blogs that provides the appeal of blog adoption in educational contexts, but for traditional book publishing the same referential system has an added benefit: where a referred resource lacks credibility or is insufficient in supporting a perspective, readers of the blogged draft can recommend more pertinent or accurate resources.

In March 2003, A-list blogger Dan Gillmor posted the introduction and first chapter of his book *Making the News* (the title was later changed to *We, the Media*). Citing the collective expertise of 1,000 editors to his publishers, Gillmor rationalized that by releasing his chapters as works in progress online, he not only was placing a degree of trust in his regular readership, but he was also managing risk in presenting material that was unfinished and open to the addition of new material on the recommendation of those readers. Designed as an editing technique, the process was apparently useful for Gillmor, as he notes in a series of interviews.[8] The blogediting of chapters allowed for correction of minor errors and, perhaps more important, expansion on ideas with relevant examples and feedback on content. There was, of course, the additional benefit of pre-marketing the book, and in spite of the fact that the entire book is now available online,[9] sales of the printed edition are strong, and the text has spawned its own blog[10] for future editions and updates.

This mimics the previously noted Rheingold *Smart Mobs* technique of extending and expanding upon the original text for future editions, but the process of blogediting is gaining credibility as a serious production mode for the traditional publishing industry, whether the works being blogedited are publicly accessible or exist as closed forums for a few key peers. In an age where time-to-market is a crucial aspect of competitive advantage, some publishing houses are recognizing that the inherent dynamism of blogging is uniquely suited to the process of book writing. And the value of a book-related blog doesn't stop at providing content for future editions of a text. Where an author is effectively untried in traditional publishing, a blog can cheaply and efficiently establish a level of subject-oriented expertise that would otherwise cost publishing houses a great deal of networking and publicity to generate.

Trialing New Authors

Brent Sampson is the CEO of Outskirt Press Book Publishing, a company that specializes in assisting new authors to self-publish their works. The company is one of many emergent players in the traditional book publishing sector, challenging the domination of mainstream publishing houses. Providing solutions to editing, printing, and distribution, these companies are responding to a perceived need in the marketplace to trial new authors and satisfy the needs of otherwise frustrated aspirants. But these self-publishing companies cannot and do not guarantee a market for the works they produce. Indeed, it is clear that the vast majority of self-published works cost more to the author than they will ever recoup in sales. However, those authors who produce content of a sufficient quality and scope to attract return on investment for published works are more inclined to be those who have honed their skills through blogging techniques. Sampson's citation began this chapter, and while he was not necessarily referring to blogs as a means of self-promotion, the use of blogs as a mechanism to practice writing and generate a perception of expertise in a field is a simple technique publishers can apply to new authors.

But the same technique can also be used by mainstream publishing houses. Using the blog as a means of trialing new authors achieves three primary functions. First, it forces authors to practice their craft by producing writing on a regular basis, and learning about how to attract new readership. Second, it forces authors to consider the feedback and free advice offered by a global audience, responding to issues raised in blog posts. Finally, it acts as a useful risk management tool in assessing possible sales of a title.

This last aspect—generating sales—is covered in more detail below, but it is a curious feature of digital-to-print production. As blog readers are attracted to the (largely self-proclaimed) expertise of a blog author, they positively reinforce that expertise through cross-referencing and trackbacks, as well as through contributing comments to blog entries. This process of referencing can then drive sales of the work by increasing exposure to the author's writing. As readers develop a relationship with the content produced by the blogging author-on-probation, there develops an emotional investment in the *author* (as opposed to the work). And the experience of several bloggers who have later released printed works (Lawrence Lessig,[11] Dan Gillmor, and Cory Doctorow[12] among others) has indicated that blog readers tend to be more inclined to purchase a work by an author with whom they feel they have developed a rapport.

Blogs therefore represent a useful means of exposing an author to the practice and process of writing and provide for the printing house a couple of useful risk minimization benefits on the side.

Publishing Operations and Blogs

The development of corporate blogs is worthy of a chapter on its own, but in the publishing sector, corporate operations blogs have the capacity to break open the traditional practices for communication between a publishing house, its suppliers and affiliates, and the reading (and paying) public. The barrier to corporate blogging adoption lies not, as is often cited, in the legal impediments to blogging online or behind the veil of an extranet, but in behavioral modification among stakeholders in the publishing sector. Search engine manager P.J. Fusco laments the speed at which corporate players tend to adopt the communication tools of the Internet:

> It took about five minutes for e-mail to emerge as an indispensable personal communications tool. It took about a decade for it to become an indispensable communications tool for corporations, generally after an outside consultant recommended it.[13]

While blogs may not have quite the same direct value as email, their function as human-driven filters for information retrieval online is clear: they are often more effective guides to special interest content online than semantic taxonomies based on keywords and concepts used by search engines. Humanization of linking then imbues blogs with an indispensability for knowledge access approaching that which email enjoys as a communications tool.

But while company executives from publishing houses may still delay widespread adoption of blogging as an editing tool, and while they may continue to wax lyrical about the great value of printed works, the same executives simply cannot ignore the use of blogs as a research tool in selling and monitoring reader responses to those printed works. Given the growth in the use of tools such as blog monitoring software (Umbria Communication's *Buzz Report*, for instance[14]), it's likely that corporates working in the publishing sector will at least be users of blogging statistics and traffic patterns, even if the use of actual corporate blogs as facilitators of editing and marketing devices for publications lags somewhat behind.

Blogging statistics and traffic patterns supply for traditional publishing houses a useful analysis of trends in audience content consumption, and provide key indicators of changing interests in subject areas. The *Technorati State of the Blogosphere Report*[15] presents an ongoing study of the growing influence of blogs as vehicles for information dissemination and filtration. As *Technorati* CEO David Sifry notes in the August 2005 report on blog influence, link behavior research is beginning to show that links to blogs are overtaking mainstream media as sources for content discussion. Traditional publishers need to take such behavioral changes into account in order to identify possible avenues for marketing titles as well as gauging the topics of interest to readers.

Further, links from such influential blogs are likely to drive as much traffic to publishers' sites or book sites as mainstream media may have done in the past in review columns. Matching discussions with source material such as the facility provided by groups like *PubSub*[16] can provide traditional publishing houses with updated and clear indications of the growing influence of blogs and the conversations arising from original works. The Massachusetts Institute of Technology-based *Blogdex*[17] scours blogs for the sites to which bloggers link in the context of their posts, thus providing a useful guide to "what's hot" in the blogosphere. All these sources represent for publishers new and zero-cost environmental scanning devices and indicators of changes in the traditional print reading marketplace, because in spite of claims to the contrary, bloggers tend to be traditional print readers.

Corporate blogging can also assist in providing efficiencies in supply chain management for the publishing sector. Desktop publishing solutions grew from the need to effectively manage the logistics of print publishing and reduce operational risk. As on-demand publishing has grown, the need for effective communications between publishing houses and their suppliers and distribution networks has mirrored that growth and placed pressure on all players in delivering titles to markets. Intra-organizational blogs operating not just in the publishing sector, but across all just-in-time industries, are beginning to emerge as a means of tracking processes in a long and complex supply chain. Blog entry syndication (Really Simple Syndication, or RSS) can inform all stakeholders in a supply chain of issues and fulfillment of orders at least as effectively as any standard ordering system. Commentator Michael Singer has argued that these intra-organizational blogs are competing with standard business communication techniques for generating operational efficiencies and marketing:

> the rise of RSS has some asking whether online businesses need to rethink their marketing strategies, when employees themselves are distributing corporate info and branding just as effectively—if not more—than normal marketing practices.[18]

Rather than retailers having to rely on their immediate suppliers for information about the availability of goods delivered along the supply chain, blogs can be used as part of an ordering process in corporate intranets to record individual order fulfillment and possible aspects impacting on the delivery of goods, as well as to provide opportunities for queries to be placed further along the supply chain. An advanced form of just-in-time servicing, the blog-ordering technique is uniquely suited to traditional publishing because content creation is dependent not merely on the capacity of printers, but on the vagaries of authorial productivity. For instance, where a publisher's capacity to fulfill retail orders for books is dependent on backorders of binding supplies, and these supplies have been affected by shipping delays on raw materials, a blog can be used not only to register such delays, but to discuss and resolve alternative solutions (contingency planning). In such cases the blog becomes an advanced form of project management tool, identifying weaknesses in the supply chain but also facilitating the kinds of discussions that need to take place in order to resolve problems. In effect, the negotiation aspects of blogs can become an innovative form of creative problem-solving.

Finally, blogs used as a marketing facility for published titles represent the most advanced and visible form of corporate blogging in traditional publishing. In June 2005, publishing house Doubleday released a fictional work by John Twelve Hawks entitled *The Traveler* and launched an elaborate series of "alternate reality" Websites to promote the work, including a *Geocities* personal blog for one of the main characters. This fictional blogging-as-marketing technique represents one of the best opportunities for publishers to engage an audience in the story being told by a fictional title without betraying the integrity of the written work, particularly where authors or endorsed subordinates are prepared to keep up posts over an extended period of time after the initial release of the book. As a source of traffic to company Websites and book sales at online Web stores, blogs as marketing devices have been shown to be remarkably successful in providing new leads. In the case of Doubleday's *The Traveler* blog, for the eight days the blog ran, it rated among the top fifty most-linked-to sites in the *Blogdex*.

Book Publishing in the Age of Blogging

Marketing commentator Seth Godin has complained that booksellers and the publishing sector generally have forgotten how to sell. Focusing on the books that people need rather than the experience they want, the publishing sector has become too convinced of its own value, its facts, titles, subjects, and prices. What it needs to do for sustainability is to communicate hopes and dreams through stories.[19] This is precisely what blogs achieve, filling a niche gap in the marketplace for consumable content. By personalizing issues and ideas through the medium of a blog, readers of tradititional publishing are moving toward blogs not only as a means of more active engagement with the content of blogs (through commentary systems), but as a means of satisfying the desired experience of reading that books alone can no longer fulfill.

Susan Herring contends that "increasing technological integration, combined with assimilation of day-to-day uses and the corresponding need to ensure the trustworthiness of one's interlocutors, will continue to make the internet a simpler, safer, and—for better or for worse—less fascinating communication environment."[20] Herring feels that the increasing comfort being generated by the act of communicating digitally, and consuming content actively rather than passively, is going to turn content vehicles such as blogs into less fascinating and more "mainstream" publishing over time. While Herring and others are not willing to predict the death of the printed book, they are convinced that the experiential elements of blogging and other interactive communication systems will attract a growing readership. Further, unless traditional print media can use blogs in the manner adopted by Doubleday in selling John Twelve Hawks's *The Traveler,* or develop interactive spaces to enhance the static material of the "dead tree" editions, increasing numbers of book readers are likely to opt for blogs as alternatives to books for their reading pleasure, and just use books for necessary rather than enjoyable content consumption.

The argument that blogs used in the editing process can devalue the content of a "finished," traditionally published work is flawed for two primary reasons: first, because the editorial process can act either as a marketing vehicle for finished works or as a risk management process to ensure quality of content developed for an increasingly competitive book publishing market; and second, because blogs simply are not the sole instrument of mass amateurization that they have often been declared to be. As Tom Coates has noted, mass amateurization is happening *everywhere.*[21] From broadcasting and music production to political activism and medical diagnosis, emergent technologies are enabling the masses to contribute content, add details, and "star" in their own productions. So rather than devaluing traditional book publishing, blogs can actually contribute to the sustainability of an industry sector that is now competing in an environment where content creation for all manner of subjects is being developed by the huddled masses. Blogs therefore represent not a threat to traditional publishing, but a great opportunity to add value to the printed word. Fortunately for the sector, many publishers and publishing industry representatives are beginning to recognize that fact.

NOTES

1. Clay Shirky, "Weblogs and the Mass Amateurization of Publishing," *Clay Shirky's Writings about the Internet: Economics & Culture, Media & Community, Open Source,* http://www.shirky.com/writings/weblogs_publishing.html, 3 Oct. 2002 (accessed 17 July 2005).

2. See Joanne Jacobs, "The Rise of Blogs as a Product of Cybervoyeurism," *Proceedings of the Australian and New Zealand Communication Association Conference,* Gold Coast, Australia, July 2003.

3. See http://www.smartmobs.com/.

4. Marydee Ojala, "Weaving Weblogs into Knowledge Sharing and Dissemination," *Proceedings of the Nordic Conference on Information and Change,* http://www2.db.dk/NIOD/ojala.pdf, Denmark, Sep. 2004 (accessed Aug. 2005).

5. Clay Shirky, "Social Software and the Politics of Groups," *Clay Shirky's Writings about the Internet: Economics & Culture, Media & Community, Open Source,* http://shirky.com/writings/group_politics.html, 9 March 2003 (accessed 17 July 2005).

6. Scott Golder and Bernardo Huberman, "The Structure of Collaborative Tagging Systems," HP Labs, http://www.hpl.hp.com/research/idl/papers/tags/tags.pdf, 2005 (accessed 17 July 2005).

7. See S.J. Liebowitz and Stephen Margolis's original discussion of network externalities at http://wwwpub.utdallas.edu/~liebowit/palgrave/network.html.

8. See http://www.itconversations.com/shows/detail404.html.

9. See http://www.authorama.com/we-the-media-1.html.

10. See http://wethemedia.oreilly.com/.

11. Lawrence Lessig's *Free Culture* is available online at http://free-culture.org/freecontent/.

12. Cory Doctorow's science fiction novels are available online at http://www.craphound.com/novels.php.

13. P.J. Fusco, "SEM and the Corporate Blog," *ClickZ Experts Advice and Opinions,* http://www.clickz.com/experts/search/opt/article.php/3517651, 2005 (accessed 17 July 2005).

14. See http://www.umbriacom.com/buzzreport.html.

15. See http://www.technorati.com/weblog/2005/08/39.html.

16. See http://www.pubsub.com/.

17. See http://www.blogdex.net/.

18. Michael Singer, "Experts See Blogs as Marketing Killer," *e-commerceguide.com* 16 Aug. 2004, http://www.ecommerce-guide.com/solutions/customer_relations/article.php/3395651 (accessed 18 Aug. 2005).

19. Seth Godin, "Buying Books Isn't Necessary," *Publishers Weekly* 252.29 (25 July 2005), p. 86.

20. Susan Herring, "Slouching toward the Ordinary: Current Trends in Computer-Mediated Communication," *New Media & Society* 6.1 (2004), pp. 26–36.

21. Tom Coates, "Weblogs and the Mass Amateurization of (Nearly) Everything," *Plasticbag.org,* Sep. 2003, http://www.plasticbag.org/archives/2003/09/weblogs_and_the_mass_amateurisation_of_nearly_everything.shtml (accessed 17 July 2005).

CHAPTER FIVE

Can Blogging Unspin PR?

Trevor Cook

E arly blogging evangelists argued that the Public Relations (PR) profession was going to be wiped away by the emergence of blogs.[1] Yet PR practitioners, often those disillusioned with the direction in which their industry is headed, are rapidly embracing the new medium as the way to revitalize PR.

In fact, the number of active PR professionals with blogs increased about tenfold or more in the twelve months following July 2004, when several dozen early adoptors put together Global PR Blog Week 1.0,[2] an event that attracted 1,000 visitors a day for a week or more. This event was recognized as one of the first of its kind in professional services.[3]

Many of these PR bloggers are sole operatives or work in small agencies, but Richard Edelman, the head of the world's largest independent PR firm (which carries his name), is also a prominent blogger.[4] He has even hired David Weinberger, one of the authors of *The Cluetrain Manifesto*,[5] as a consultant to his firm on blogging.

Steve Rubel, the world's most successful PR blogger,[6] a genuine blog A-lister, is building a new practice in his New York PR firm (CooperKatz) that is using the new participatory media to achieve the sort of results for his clients that PR has traditionally been charged with delivering. One of his clients, Weatherbug, is now recognized as an exemplary case study on the use of blogging for PR purposes.[7]

> **Participatory (or Social) Media:** media that eliminate, or minimize, the traditional division between producers and users in the creation, distribution, and consumption of news and commentary.

We live in an ironic and cynical age. Our children are taught to deconstruct media messages, and our popular culture resonates with the idea that PR manipulation means that fact is increasingly difficult to distinguish from fiction. In addition, with the growth in recent years of dubious and often downright unethical techniques like product-placement, astro-turfing, and cash-for-comment deals, the public relations industry faces a crisis of legitimacy.[8]

> **Citizen Journalism:** generally refers to the empowerment of "ordinary" citizens to use the traditional techniques and skills to produce news, usually for publication online.

The PR industry's crisis has two dimensions. First, the end of media scarcity undermines the central platform on which much of traditional PR is constructed. In a post-scarcity age, PR needs to rethink its role and its tactics. PR does not need the media in the same way and to the same extent as in the past. Nor can we rely on the media to do our jobs for us. Media remain important, but they are tumbling from their pedestal with the rise of citizen journalism and social media.

In addition, the emerging Web 2.0[9] is generating the techniques and skills to allow people to effectively "edit" the information they choose to access and, significantly, to re-package and re-formulate that information for sharing with others.

> **Web 2.0:** describes a set of tools that make the Web more dynamic, more interactive, and more supportive of networking.

For instance, the widespread use of RSS (site syndication),[10] which allows sites to talk to each other and reduces the need for users to visit Websites, is changing the way people interact with the Web itself. Using feeds and aggregators, people have much more control over the information environment they inhabit. With Web 2.0, the Internet is becoming a more dynamic—and appealing—publishing and broadcasting environment.

Confronted with this at least partial dismantling of the old news media edifice, PR has to become more creative and more open in the way it communicates with stakeholders.

Second, PR has always been weakest in the area of communications between individuals. The broadcast age made it cheaper and more efficient to simply flood audiences with carefully researched and tested messages rather than genuinely to engage them on a "one-to-one" or "many-to-many" basis. In practical terms, the mechanisms simply did not exist to do this in a mass industrial age.

Will blogging and other forms of social media lead to a new age of open, accountable, and interactive communication in Western democratic societies? It's still too early to tell, but if it does, adventurous, reformist PR people may well be at the forefront of ushering in that new age.

Participation and the End of Media Scarcity

New media formats, particularly blogging, are participatory in a number of ways. They are participatory in the sense that anyone with access to an Internet-connected computer can now be a publisher of news, ideas, and opinions. Blogs, as easily created and edited Websites, are fulfilling one of the Internet's early and most cherished ideals—the removal of the distinction between author and reader. On the Web, Tim Berners-Lee[11] and others hoped, the power relations between producers and consumers of information would disappear. Until the emergence of blogging software, however, the difficulties involved in creating and editing Websites deterred all but the most technologically savvy.

New Technology and the Culture of Participation

The implications for traditional media of the belated emergence of the "Web as publishing environment"[12] have been widely discussed, most famously in Dan Gillmor's *We the Media*.[13] Gillmor has done a lot to popularize the notion of blogs as the platform for grassroots, or citizen, journalism. Another keen observer of this development calls bloggers "stand-alone journalists."[14] With the advent of blogging, journalism is now to be defined as something you do, no matter whether on your own blog or in a Murdoch-owned publication.

Consequently, the traditional role of the media as gateway is eroding because editorial space is now much more widely available. These more elastic media environments can now expand to provide publishing opportunities for anyone who has something they want to say. As citizens, or PR practitioners, we don't have to "pitch" our story ideas to journalists and editors and hope that we get some coverage in the daily paper or on the nightly bulletin.

Blogs are participatory in another sense, with perhaps even greater implications for PR than the eventual disappearance of the media gateway. Blogs encourage conversations to an extent that has not been possible in any other communications activity except personal interaction. PR professionals generally recognize that personal interaction is still the best communication mechanism, and all others are little more than poor substitutes or reinforcing supplements.

Traditional news media jealously guard their exclusive space. Apologies and retractions are rare and grudging; letters to the editor provide some limited opportunities for dialogue, but this part of the newspaper is also carefully orchestrated by the letters editor. Talkback radio provides significantly more opportunities for citizen participation, but again it is limited and often heavily manipulated.

On the other hand, much of the software embedded and associated with blogging pushes us in the direction of participating in communities that are continually forming and reforming and swirling around particular topics and interests. Comments, trackbacks, links and tags are all features that give blogging its special characteristics. The idea of community is also at the heart of the culture that supports blogging. Unlike traditional news media, participation is not an add-on but at the very heart of what the new "participatory" media are all about. In fact, without this culture of community, it is unlikely that blogs would be anything more than a flotsam of cheaply produced Websites.[15]

Breaking the Nexus between PR and Media Scarcity

PR emerged during the twentieth century in part as a counterweight to the news media's gateway role in public conversations. The news media play a significant role in shaping public agendas by deciding what we are—or should be—interested in and by deciding what is news and who are the newsmakers. PR exists in large part because governments and corporations are unwilling to cede control of their communications with their stakeholders to independently minded, but often hostile, media.[16]

Typical criticisms of the media revolve around their lack of expertise in particular subject areas, their drive to sensationalize, and their unwillingness to present both sides of a complex issue or to admit error when they get things wrong. While some of these shortcomings can be attributed to the commercial imperatives that drive corporate media to maximize audiences, some of it is also due to the arrogance of an oligopolistic mediascape, which too often sees itself as the (sole) arbiter of "truth." The capacity to counter this arrogance is why blogging and podcasting are sometimes described as the democratization of the media.[17]

Faced with limited opportunities for coverage, and the distortions of big media journalism, traditional PR techniques are a plausible, often successful response. If you are going to get one, or just a few, opportunities to talk through the media to your audiences, then it makes sense to focus on a small number of messages. If journalists spend most of their time looking for conflicts and inconsistencies, then it is easy to be obsessive about eliminating these wrinkles, no matter how trivial, from your story. If these oligopolistic media crave bad news and put undue emphasis on negatives, then it makes sense to try to "accentuate the positive and eliminate the negative."

If there has always been a symbiotic alignment between the media as "gateway" and the PR profession as "message controller and distributor," then an eroding of one side must lead to significant changes for the other as well.

PR with the Gateway Open

Corporations and governments in the past have tended to believe that the best decisions are made in secret. Since Watergate, if not before, one of the biggest mistakes anyone can make in PR is to cover something up. In the end, the "cover-up" is often seen by the public as worse than the original mistake. Nevertheless, many organizations still try to severely limit the flow of information into the public arena, and their PR advisers, unfortunately, are too often complicit in this flawed approach.[18]

This desire for secrecy also results partly from an inability to compete with the media in terms of getting information out to the organization's publics, and from the control journalists and editors have over the flow and presentation of information to "their" audiences.

Publish or Be Damned

With the growing popularity of blogging, this problem is being diminished. Organizations can release information on their blogs before it is in the media, and they can respond to media distortions quickly and effectively. If what they say is interesting, other bloggers will link to and comment on it, and it will get the sort of notice and attention that was only possible through news media coverage in the past.

Not only can organizations become publishers, they must do it. There is a growing public expectation that organizations will provide information and commentary in much the same way that the media do. Whereas traditional Websites were little more than brochures and e-commerce vehicles, today we are looking to organizations to use the Web as a publishing environment, in the sense of providing a continuous flow of fresh, time-critical material. Just putting up media releases, speeches, and annual reports doesn't cut it anymore. People consume the news differently.

The Internet is making the news cycle of regular editions and bulletins obsolete. The searchable Web, and millions of feeds, are separating the concept of "news" from the time-based anchor attached to it by the technological constraints of traditional media.

From Messages to Conversations[19]

With the emergence of blogging, millions of people around the world are now participating in publicly visible discussions.[20] These discussions are more like genuine human interactions than the one-way downloads of traditional news media and PR. People much prefer the "cut-and-thrust" and the unscripted nature of conversations. They also relate to identifiable personalities.

Much of branding is about giving an organization a clear and consistent "personality" that customers can feel a connection to, or relationship with. But relationships are created between specific, concrete individuals. It is the specific characteristics of an individual that we appreciate, not generalizations. You can't have a beer with a logo; you can't chat over the back fence with a brand descriptor.

Blogs emphasize individuality. If you read someone's blog for a while, their personality shines through, and it is possible to appreciate the complexity that makes them a real person. Once that happens, conversation comes more naturally. Messaging involves simplicity (dumbing-down) and repetition, whereas conversations allow people to explore ideas in depth and tease out issues. With messaging, feedback is usually stilted and formal—the suggestion box, the focus group, the satisfaction survey, and so on. With conversations, the feedback can be more layered, subtle, and immediate.

While blogs cannot match the power of being in the same room as someone, they do have some distinct advantages over face-to-face communications. Blog conversations are not bound by the normal constraints of time and geography. A conversation can be started by a blogger in one part of the world and spread across the Internet, involving people from anywhere and everywhere. Blog conversations can also be more democratic, as many people find it easier to comment from the relative security of their keyboards rather than having to stick their hands up in a room full of people.[21]

What's more, the conversations on, and between,[22] blogs can be read, googled, and bookmarked, and they become part of a transparent, permanently accessible, public record that continues to shape public opinion long after newspapers have become fish and chips wrappers.

It should be noted here, however, that blogging is not the first medium to harness the power of participation. Talkback radio became a politically powerful medium in the United States and Australia in the 1990s by being able to connect with people in ways that newspapers and magazines cannot rival. The difference is that bloggers can participate in conversations on their own terms, and their participation is not controlled by a talkback host. Talkback versus blogging is like centrally controlled participation compared to distributed involvement.

Nevertheless, PR people should see blogs and bloggers as the talkback shows and hosts of the Internet Age: any blogger who has an audience of enthusiastic participants (either through comments, trackbacks, or links) is in a position to shape opinions and shift market behaviors just by having these public conversations.

Old PR tends to be episodic. We put out a media release, push it to the media, count the column inches, and plan our next effort. Companies often think they are doing well if they put out a couple of releases each month and are even happier if one of these gets picked up by a major media outlet. Yet, this smacks of the horse-and-buggy age of PR compared to what can be done online.

Blogging makes PR a much more free-flowing activity. It is a dialogue that does not depend on regular "announcements" to meet the media's time-dependent definition of news. The conversation might get googled by someone a few months later who is looking for information on your organization. Each time that happens, the contents of the conversation again become "news" for that person at that time.

PR needs to think of communications as something that is flowing all the time, something that is happening all the time, whether we like it or not. Paradoxically, this free-flow style of communications might prove a lot easier to maintain than manufacturing news for the media, especially for clients and on subjects that the media do not find "sexy" or appealing to sufficiently large (paying) audiences.

Talking in Depth

A common complaint about the media is their propensity to dumb down. Of course, this is largely the inevitable consequence of commercial strategies that depend on audience size. The easiest way to reach a bigger audience is to produce content that is more broadly acceptable. The result is, as Evelyn Waugh put it in his 1930s satire *Scoop*, that newspapers are written for people who aren't interested in anything—the idea being that anyone with anything more than a cursory interest in a subject is not going to turn to the media for information about it.

They might now, however, turn to a blog written by a specialist for specialists. In fact, many of the most successful blogs fill just this sort of gap in our traditional media environment. Something like *Digital Photography*,[23] maintained by Melbourne-based blogger Darren Rowse,[24] is an excellent example. Darren maintains a stable of about twenty blogs which together made him a phenomenal US$15,000 in May 2005. *Digital Photography* is one of two or three blogs that bring him most of that revenue. Podcasts fill a similar role in the existing radio landscape: *The Podcast Network*,[25] started in February 2005 by two Australian entrepreneurs, is on the way to assembling a stable of about one hundred shows, many of which cover subjects that are not commercially viable in radio, or cover them in a depth and with a frequency that is not viable elsewhere.

There are also increasing numbers of academics, activists, and professional services providers who are blogging for audiences that are smaller and more expert than the audiences for more traditional vehicles, like op-ed pieces.

A classic PR strategy is to become the expert, or thought leader, in a particular subject area. Until now this has been done through the fairly laborious efforts involved in writing books, making seminar presentations, and, it is hoped, a steady flow of media interviews with journalists keen for a quote from an acknowledged expert. Keeping a blog is a much cheaper, easier, and more accessible way of achieving the same outcome. Doubtless, it won't be too many years until someone who does not maintain a blog on a subject just cannot be credible as an expert in that area.

PR Must Drive Cultural Change

If blogging is to change the way in which corporations communicate internally and externally, then it will be PR blogging evangelists who will be sounding the clarions and showing the way. Make no mistake: it will be a tough job, and much depends on the skill PR practitioners can bring to the task of guiding their organizations through the minefields involved in major cultural change.

By March 2005, Jeremy Wright[26] claimed that at least 10 percent, and perhaps 20 percent, of Fortune 500 companies were using blogs in some way. He posted a list of forty of these companies, which he had identified as part of his research for a book on business blogging. So far, these major corporations have only dipped their toes into the blogosphere, and it is unlikely that many, if any, have changed their corporate cultures much as a result.

Corporate communications presents a disembodied "voice" to the external world. Where necessary, a spokesperson is deployed to convey the previously crafted messages. The emphasis is on avoiding mistakes and minimizing untidiness. The messages are usually designed to be safe and to chloroform discussion rather than fuel it. The messages are all good news about the organization, and in line with sound PR practice; people also avoid talking about competitors. Talking about the opposition simply gives them "oxygen" and wastes precious editorial space.

PR professionals and their corporate bosses generally believe that it is the media's job to find flaws in these messages, and it is their job to "accentuate the positive and eliminate the negative." But if we are to move into the new media space (as we must now that the gateway role is eroding), then we have to accept some of the roles and responsibilities traditionally associated with good journalism. That means emphasizing qualities like fairness, balance, accuracy, and integrity in our own materials rather than slanted, hyperbolic advocacy that ultimately relies on the third-party endorsement of a trusted media brand for its credibility.

It is a big stretch for corporates to use their Websites, or other platforms, to talk about their own problems or give credit where it is due to competitors.

Successful corporate bloggers, like Microsoft's Robert Scoble,[27] have to maintain a certain distance from their organizations if they are to be seen as credible. If Scoble were to duck embarrassing issues too often, or fail to mention the success of other corporations, and just stick to a relentless stream of positive stories about Microsoft, his audience's reaction would be savage. He makes a genuine effort to be fair, but he gets hammered by critics every time he says something positive about Microsoft. It is impossible for someone to blog about their own organization and to be totally impartial, but the blogosphere will always be pushing bloggers in that direction. Unless we move in that direction, we will not get the benefits on offer through blogging.

As well as helping their bosses to think and act more like good journalists, PR people are going to have to show the way in reforming the current abuse of the English language that occurs in media releases and all other corporate communications. This linguistic abuse is not just irritating; it also has negative consequences for the health of our democracy and the credibility of our materials.[28] If we are to carry on conversations with real people, then we are going to have to talk in real language.

Finally, a further cultural issue that PR needs to grapple with is the extent to which an organization can offer a number of spokespeople (bloggers) and let them talk about the company without the whole exercise degenerating into chaos and confusion. At one level, every organization needs to make authoritative statements about its strategy and policies, and it needs to avoid confusing stakeholders by putting out a range of conflicting statements on important (particularly market-sensitive) issues. At the same time, corporations do need to be more flexible and to allow their stakeholders and employees to participate in these powerful public discussions that can do so much for the organization and its standing in the marketplace. Getting the balance and the rules right will test the skills of the most proficient PR operatives.

Change Is on the Way

Good PR practitioners have always seen themselves as the champions of better engagement between organizations (or clients) and their "publics." In the process, they resist the organizational bias toward secrecy (or "flying under the radar") and represent stakeholders to often insular management teams. Up until now, the news media scarcity problem has provided second-rate practitioners with a plausible rationale for understanding PR as the practice of message control and obfuscation. That rationale is rapidly disappearing.

Blogging can be an answer, even perhaps *the* answer to the crises facing PR, if the industry gets behind it and shows the corporate world how to use the new media to great advantage. Along the way, the cultural changes needed to take full advantage of the new technologies can also change corporate communications for the better. When we look back on the history of PR, we might one day say that blogging arrived just in time.

NOTES

1. See Jay Rosen, interviewed by Steve Rubel, "PR Needs to Stand for Real Transparency," http://www.globalprblogweek.com/archives/jay_rosen_pr_needs_t.php (accessed 26 Sept. 2005). See also Darren Barefoot, "Sorry, PR Ain't Dead," http://www.darrenbarefoot.com /archives/001202.html, (accessed 26 Sept. 2005) for an alternate view.

2. Global PR Blog Week 1.0 papers and comments are available at http://www.globalprblogweek.com/.

3. Comment made by *MarketingSherpa* site, http://www.globalprblogweek.com/archives/ blog_and_media_cover.php (accessed 26 Sep. 2005).

4. Richard Edelman's blog is called *Speak Up* and can be found at http://www.edelman.com/ speak_up/blog/.

5. Frederick Levine, Christopher Locke, David Searles, and David Weinberger, *The Cluetrain Manifesto: The End of Business as Usual* (New York: Perseus, 2000).

6. Rubel's blog is called *Micro Persuasion* and can be found at http://www.micropersuasion.com/.

7. See the *Technodaddy Abides* blog, http://www.technodaddy.com/wp/2005/06/25/steve-rubel-tomorrows-public-relations/ (accessed 26 Sept. 2005).

8. Trevor Cook, "Buying Influence: Cash for Comment Scandals and the Consequences of PR Gone Feral," *The Walkley Magazine*, Feb. 2005, also at http://trevorcook.typepad.com/ weblog/2005/03/buying_influenc.html (accessed 26 Sept. 2005).

9. Jared M. Spool, *Web 2.0: The Power behind the Hype*, http://www.uie.com/events/uiconf/ articles/web_2_power/ (accessed 26 Sep. 2005).

10. See http://en.wikipedia.org/wiki/RSS_(protocol) for definitions, usage, and history.

11. Berners-Lee built the world's first Website (1991) and is acknowledged as the creator of the World Wide Web; more at http://en.wikipedia.org/wiki/Tim_Berners-Lee.

12. Dave Winer, "The Web Is a Writing Environment," *Davenet*, 17 Apr. 2001, http://davenet.scripting.com/2001/04/17/theWebIsAWritingEnvironment (accessed 25 Sep. 2005).

13. Dan Gillmor, *We the Media: Grassroots Journalism by the People for the People* (Sebastopol: O'Reilly, 2004).

14. Jay Rosen, journalism professor at NYU and author of the *PressThink* blog, http://journalism.nyu.edu/pubzone/weblogs/pressthink/.

15. Trevor Cook, "Politics and Community-Building Online," *New Matilda*, 1 Dec. 2004, also at http://trevorcook.typepad.com/election/2004/12/politics_and_co.html (accessed 25 Sept. 2005).

16. Trevor Cook, "A New Spin on Blogging," *The Walkley Magazine*, Oct. 2004, also at http://www.podcastingnews.com/archives/2005/09/nova_podcast_lo_1.html (accessed 25 Sep. 2005).

17. Shane Bowman and Chris Willis, *We Media: How Audiences Are Shaping the Future of News and Information*. See in particular chapter 5, "Implications for Media and Journalism," http://www.hypergene.net/wemedia/weblog.php?id=P41 (accessed on 25 Sept. 2005).

18. Gerry McCusker, *Talespin: Public Relations Disasters—Inside Stories and Lessons Learnt* (London: Kogan Page, 2004).

19. The idea of markets as conversations and its relevance to the blogosphere was popularized in Levine *et al.*, *The Cluetrain Manifesto: The End of Business as Usual.*

20. David Weinberger, "From Public Relations to Public Relationships," *Joho the Blog*, 16 July 2005, http://www.hyperorg.com/blogger/mtarchive/004240.html (accessed 25 Sept. 2005).

21. Julie Leung, sessional leader, "Emotional Life," *Bloggercon3*, Stanford University, 6 Nov. 2004, MP3 audio file available at http://www.itconversations.com/shows/detail286.html (accessed 25 Sept. 2005).

22. In a discussion at Blogtalk Downunder in Sydney in May 2005, Mark Bernstein made the point that the real communication in the blogosphere occurs between blogs, not as comments on individual blogs; see http://possibleworlds.blogs.com/blogsperiment/ (accessed 28 Sep. 2005).

23. The *Digital Photography* blog is available at http://www.livingroom.org.au/photolog/ (accessed 25 Sept. 2005).

24. Darren Rowse announced his May 2005 earning on his *problogger* site; see http://www.problogger.net/archives/2005/07/12/earning-milestones/ (accessed 27 Sept. 2005).

25. *The Podcast Network*, http://www.thepodcastnetwork.com/ (accessed 27 Sept. 2005).

26. Jeremy Wright, *Ensight*, http://www.ensight.org/archives/2005/03/07/how-many-fortune-500s-blogging/ (accessed 27 Sep. 2005).

27. Scoble's blog, *Scobleizer*, can be found at http://scoble.weblogs.com/ (accessed 27 Sept. 2005).

28. Don Watson, *Death Sentences: How Cliches, Weasel Words and Management-Speak are Strangling Public Language* (New York: Gotham, 2005); George Orwell, *Politics and the English Language*, 1946, http://www.resort.com/~prime8/Orwell/patee.html (accessed 27 Sept. 2005).

Blogs in Business:
Using Blogs behind the Firewall

Suw Charman

U ntil recently, internal business blogs have been *terra incognita* for anyone not directly involved in an internal blogging project. There has been a large degree of reluctance to open up and allow researchers access to internal blogs, possibly for fear that by discussing such projects the company may lose competitive advantage, or because of a general reluctance to talk about unproven technology.

Because they are behind the firewall, they cannot be observed *en masse*, but, just as scientists say that the universe is full of dark matter, so business is increasingly filled with these "dark blogs." As blogging becomes more mainstream, more examples of dark blogs are coming to light, more research is being done, and it is starting to become clear how this technology works in a firewalled environment.

> **Dark Blogs:**
> blogs not accessible to the public because they are behind a firewall or password.

Why Use Blogs?

Blogs are frequently portrayed in the media as personal diaries, the contents of which are either trivial or libelous. While it is true that the majority of public blogs are of interest only to the friends and family of the blogger, the depiction of blogs as diaries distracts from their real nature by conflating content and tool.

When viewed as a simple, flexible, cost-effective, lightweight content management system (CMS) with additional social networking capabilities, the potential applications for blogs in business become clearer: a blog can be an event log, a newsletter, or a project management tool, depending on how it is used.

The reasons for using blogs are many. Blogs have a flexible format that allows for unstructured items of varying lengths and encourages simple categorization of data, are easy to implement and use, require little or no training or knowledge of HTML, can be integrated with existing systems, and are low-cost.

The Problem with Email

> **Occupational Spam:**
> business email sent internally that has no relevance to the recipient's job.

Email is increasingly "broken" in the corporate environment. Workers are being subjected to what is often called "occupational" spam—emails sent by colleagues that have no relevance to the recipient's job. Reports show that workers are getting an average of 97 emails a day,[1] and that around 34 percent of them are occupational spam.[2] Employees are now spending too much time clearing out their inboxes, and the sheer volume of email puts them under undue pressure as they try to keep on top of it all.

But the problem with email goes beyond spam. Email is primarily a one-to-one medium that breaks down when used for group communications and collaboration. People are accidentally left off the CC list, and the conversation splinters, diluting the "group memory" and making version control of attached documents difficult. Eventually, the entire email thread gets split across individual archives or deleted when users reach their inbox size limit. This lack of persistence means that knowledge and information is permanently lost to both the company and the individual.

> **RSS:**
> RSS (Really Simple Syndication, or Rich Site Summary) is a standard XML format used to share blog headlines and other pointers to Web content.

Blogs and RSS (Really Simple Syndication, a content packaging protocol) can be used to replace email for non-urgent communications such as newsletters, updates, or notifications. By posting such information on a blog instead of by email, the volume of email is lowered, the signal-to-noise ratio in employees' inboxes is improved, and archives become open and easily searched.

Cultural Sea Changes

There are two main differences between personal and business blogs:

- Business blogs have a tendency to be factual and outward facing, using a linklog style that points to interesting or relevant information on the Internet. There is less discussion of the issues people are facing in their jobs than one might expect from the culture of openness we observe in public blogs, so they are rarely used as a way of soliciting help, information, or support.

  > **Linklog:**
  > a blog comprised predominantly of links to other blogs, with little or no commentary.

- While an informal writing style is often employed by business bloggers, internal blogs are not usually the same center for free-flowing conversation that personal blogs can be. Instead of leaving a comment, readers prefer to respond by phoning or emailing the author. This privatizes the discussion, removes it from the intranet domain, and, in doing so, misses out on the opportunity to add value to the original post.

The open, conversational nature of blogs is a cultural sea change that some may feel uncomfortable with. Business communications have become highly formalized, to the point of having an established business lexicon that frequently bears little relationship to reality.[3] Some companies have also developed a strong etiquette that discourages employees from expressing privately held opinions.

Blogging breaks down communications barriers, providing employees with a forum to talk among themselves and to disagree with the company or company policies—activities that management can find threatening. Such behavior is seen as highly disruptive, and the unenlightened manager's response is to try to stamp out such discussions rather than see them as potentially valuable and constructive feedback.

Increased insecurity in the job market has resulted in an office culture where long hours are normal and employees feel the need to be seen not only to be working at all times, but also to be eschewing coffee and lunch breaks. In a company with an over-developed work ethic, work can be split into two types: "visible" work, which includes things like meetings, writing reports, or replying to emails; and "invisible" work, such as research, discussion, and reflection.

> **Visible Work:**
> tasks that require noticeable effort and activity.
>
> **Invisible Work:**
> tasks that require thought and reflection, but no noticeable activity.

Like all writing, blogging begins with reading, but browsing the Internet is often seen as an unnecessary activity in a business context. Indeed, there is anecdotal evidence that, in some companies, even having a browser window open can result in castigation, despite the fact that a browser is required to access the company's own intranet. Thus develops a tension between the blogger, whose job relies on doing invisible work, and management and colleagues, who expect the blogger to do only visible work.

Normative Forces

In general, though, fears about the publication of inappropriate content are unfounded. While the opportunity for misuse of dark blogs is vast, in actual fact most people are sensible enough not to post anything they may later regret or that may get them fired, in the same way that people are generally careful about what they say when the boss might be listening.

If anything, the very thing that is feared—the open nature of a business blog and the fact that what you write is visible to the rest of the company, including your subordinates and superiors—creates a self-regulating system. Dark bloggers feel wary of what they say on the intranet, not necessarily out of fear for their jobs or of humiliation, but out of a sharply honed (some may say over-developed) sense of what is appropriate and a desire to err on the side of caution. In some cases, this may actually inhibit bloggers who otherwise would make a valuable contribution, but it takes time, effort, and trust to break down habits of circumspection ingrained by years of exposure to subtle company censure. Simply giving someone the tools to communicate does not guarantee that they will.

The fact that business blogs appear to always be attributed—there are currently no examples of anonymous dark blogs known to the author—also acts to help bloggers behave in a responsible manner, because it is clear that this material is neither anonymous nor private.

Conversely, email creates an illusion of privacy that gets people into trouble because it is too easy to presume privacy and forget that not only can an email be forwarded, but that it is also being stored by the company for legal reasons. This can lead people to write email that they would never send if they knew it was going to be read by anyone other than the recipient.

Dark blogs may be behind the firewall, but they are clearly not private, and established social norms constrain users, encouraging responsible behaviors.

What Are Dark Blogs Good For?

As with the empty notebook, blogs can be put to almost any use imaginable, but usage falls into three main categories.

1. Individual blogs. Written by one person as a way of taking notes and/or communicating with a small constituency of co-workers.
2. Group blogs. Written by a group of people, these collaborative blogs can be used for collecting information gathered by a team or department, or for workers who need to communicate between shifts or time zones. The readers are generally also the writers.
3. Company blogs. Written by an individual or group, these blogs are more broadcast-like, intended for everyone within a department or company. The readers are generally not the writers.

1. Individual Blogs: Personal Knowledge Management

Possibly the most popular use for dark blogs is personal knowledge management, an increasingly important task for anyone working with large amounts of information that they have to track, store, or report on.

Tristan Leaver, Head of Business Development for *The Guardian*, started a blog initially just to store snippets of information, interesting URLs, details of meetings and phone calls, and to aggregate data coming in from other tools such as *Del.icio.us*, *Google News Alerts*, or *Technorati*. After a while, he discovered that a secondary benefit to blogging was that it made it easier to disseminate information to the rest of his team and cut out the need to write monthly progress reports.

> A blog is a good way of keeping people who don't sit together up to date with what's happening and to provide a [point of] reference, so if you want to go back and find out what happened, you can. And at the end of the month, [instead of writing a report] I can print the details of everyone I saw, what my thinking was, and why we're doing what we're doing.
> Previously, if I had a meeting with a supplier, I'd have to write it all in an email and send it to someone. And then I'd have to take that email, cut and paste it into a Word document and save it somewhere, or put it on our contact management system. Now I just put it on the blog and if I think anyone else is interested I just send them the URL.[4]

2. Group Blogs: Event Logging

Businesses where workers are employed on a shift basis, and where issues persist across shifts, require better-than-average communication. Ensuring that each shift knows what happened on the watch before and can communicate with those coming into work after them is essential to the smooth running of the business.

Disney/ABC Cable Networks is responsible for the broadcast of the *Disney Channel*, *Toon Disney*, *SoapNet*, and *ABC Family*, working 24 hours a day, 365 days a year and with over 100 staff working three major shifts.[5] It is crucial that information is communicated efficiently and that archives of past logged events are accessible and searchable.

When Disney first started broadcasting, any problems were logged in a book, and when staff changed shifts they would consult the book to see what had happened. Eventually, this process was moved to a specialized, but functionally limited, database which logged the entries and produced email alerts, but which could not be searched or edited.

Mike Pusateri, Vice President of Engineering, decided to replace the old database with blogs and migrate some 9,000 entries into the new installation. The categorization remained the same as in the old system, but the new blog software added search, comments, and archives, immediately creating value in the old log entries.

A combination of the RSS produced by the blogs and FeedDemon, an RSS reader that integrates with Microsoft Outlook, allowed people to access the shift log via email. Because the old system had created an email alert for every new database entry, regardless of relevance, using RSS feeds for each category actually reduced the volume of email, since individuals subscribe only to the RSS feeds that they need.

The blogs also create a shared archive, where everyone can see the history of any given problem and nothing is hidden from view, a feature entirely absent from both the old database and from any email-based communications.

3. Company Blog: Knowledge Sharing

> **Knowledge Management (KM):**
> gathering, organizing, sharing, and analyzing data, information, and knowledge.

Sharing data, information, and knowledge is widely recognized now as an essential business practice. In fact, an entire industry has been built around creating software and processes for doing this, so-called "knowledge management" (KM) tools. Highly complex KM systems which required expensive customization for each installation promised to unlock the power of all the knowledge hidden away in employees' minds, yet KM projects often failed to deliver.

One of the reasons for this is that knowledge is actually very hard to manage, and most KM systems focused more on data and information (in the form of documents) than actual knowledge.

Blogs, however, allow for a personal narrative which, along with discussion between blogger and readers, can turn data or information into knowledge. By providing simple categorization, search, and linking facilities, blogs create a web of knowledge that can be easily accessed on a convenient, *ad hoc* basis.

One unnamed pharmaceutical company in Europe decided to use blogs to share competitive intelligence (CI) information.[6] "I wanted to put in place a corporate repository for competitive intelligence material, a way to capture and share a knowledge base of relevant information about our competitive environment," said the CIO.

The project was supported from the CEO down and was formally planned and implemented by a CI team working with an enterprise blogging software vendor to create a set of blogs suitable for an organization where access to the information had to be restricted to clearly defined groups and content vetted by a dedicated editorial team.

Blogs as Website Management Tools

Blogs are, in essence, a lightweight content management system, so it is not surprising to see them being used as such.

In the early days of the Web, when businesses were just starting to realize the usefulness of having internal Websites, many intranets were built as static HTML sites by a central Web development team. Then came vendor-driven CMS tools such as Documentum, which allowed a degree of "self-service" for those employees who were given permission to add content.

Blogs are now being used to replace these large-scale CMS tools completely, and at a fraction of the cost. Some blogging tools can be highly customized, so it is possible to use the same system to create both dark blogs and an intranet that, on the surface of it, looks much like a traditional Website. By using blog software, it is possible to open up creation of the intranet to any employee with a log-on, but without requiring them to grapple with a complex workflow system, thus cutting down on intranet maintenance overheads.

Blog Implementation: Top Down vs. the Trojan Mouse

It appears that there are two main ways in which blogs are implemented in an internal business context: as a top-down, management-approved project, or as a Trojan Mouse.

A Trojan Mouse is a small IT project started by an individual with access to hardware that is not under the control of the IT department, thus allowing the project to develop out of sight of management. Once the software is installed, usage spreads beyond the originator by word of mouth, rather than through any sort of promotional campaign.

> **Trojan Mouse:** small project or piece of software implemented or installed without managerial approval or knowledge.

Eventually, the Trojan Mouse is discovered, and one of two things can occur: either the software proves too valuable, or the users too many, for the project to be shut down; or the IT department takes exception to unauthorized software and removes it. It seems that, in general, Trojan Mice are successful at surviving once discovered, because by the time they are located, they have proven their benefits to a wide enough audience, often including senior staff.

The other main route for blogs to find their way into business is for a senior manager to be convinced of their usefulness and then instigate an official blogging project.

Trojan Mice tend to grow very organically, changing nature with the changing demands of the users, whereas successful official blogging projects are usually carefully planned (although the wise blog manager also has the flexibility to allow for emergent behaviors).

A Brief How-To

Implementing any blogging project, top-down or Trojan Mouse, requires a degree of thought and consideration. Only users who are already familiar with blogging seem to take to blogs on a free-for-all basis. (Most users require a more thoughtful approach.) Here we run through the main points to be considered when planning a blogging project.

Have a Reason

Identify a real business need. What problems have you identified that a blog can address? Blogs that serve no obvious purpose but that have just been put up "for the sake of it" tend to fail. Not all problems have a blog-shaped solution, so consider precisely how a blog would work in your business environment.

Consider the Way People Work

Adoption is frequently cited as a concern among those who are considering implementing dark blogs, but if your blog solves a problem and fits into the way people already work, then they will use it.

It's important to realize that blogging takes time and effort, so there has to be a significant reward. Perhaps blogging replaces the compilation and distribution of monthly reports, makes logging and finding information easier, or cuts down on the amount of time the user spends processing email. If users clearly understand the benefits, they will be happier to learn how to use the new software.

Blog projects that benefit the company but not the user will meet resistance. This has been clearly demonstrated by KM projects which cost hundreds of thousands of pounds to implement but which were ignored by the very people who were supposed to populate them with documents. If users see no reward for putting in the extra effort required to populate a KM system with content, they will find ways not to do so.

Integrate

What business systems do you have with which your blogs could be integrated? For example, integration with LDAP corporate directories to create a single log-on, or with email so that users can email entries to their blog, both lower the barrier to entry.

Consider, too, how your users are going to access the blog. Although the obvious way to read blogs is through a browser, if your users potentially have a lot of blogs to keep track of, then it makes more sense for them to use RSS. A desktop RSS reader can help employees by creating a separate environment for reading blogs, thus avoiding the "noise" of a busy email inbox.

If users aren't comfortable with specialist applications, use email as a bridging technology between new and existing environments—RSS feeds can be integrated with email so that they appear in the email client in the same way that newsgroup headlines or email do. Additionally, email alerts or daily digests of recent headlines are a good way to gradually draw reluctant users into blogging.

Provide Support, Not Training

Training is, in general, not very popular. Most people prefer not to have to deal with a 300-page user manual or the three-day course. Blogging tools are very simple and just don't need this level of training. Instead, provide one- or two-page cheat sheets and back them up with on-demand support. Where required, training should be short, simple, and informal.

Beware Knowledge Hoarders

Modern business culture puts strong emphasis on what you know, and many people believe that they need to hoard their knowledge in order to retain status and value within their company. For these people, sharing their knowledge on a blog is anathema, and they will do everything they can to avoid doing so. The change from hoarding to sharing is a big cultural shift, and again the benefits of opening up need to be made clear so that knowledge hoarders realize that their value actually increases the more they share.

Allow for Emergent Behaviors

Blogs are very flexible tools, so don't be surprised if people start using them in ways you hadn't anticipated. In fact, if they do, then this is a good thing—it means that they've accepted the new technology, taken it to heart, and are using it in the way in which they feel most comfortable. Embrace emergent behaviors and be open to sharing them with other users.

The Future

As companies realize that the benefits of blogging go beyond the financial, and that this so-called "disruptive" technology can in fact be highly constructive, there will be more adoption of blogging behind firewalls. Eventually, blogs will stop being new and different, and will be just another tool in the businessman's kit. No one now talks of the telephone as if it were something unusual or remarkable, and blogs will soon be just as common.

As adoption increases, we will see a development of dark blogging technology to reflect behind the firewall the functionality that is common in the public blogosphere. Tags, trackbacks, social bookmarking, collaborative annotation: all these tools and techniques will eventually be replicated behind the firewall, and then we will truly see movement toward Intranet 2.0.

NOTES

1. Radicati Group & Mirapoint, "End User Study on Email Hygiene," April 2005, http://www.radicati.com/brochure.asp?id=75 (accessed 3 Nov. 2005).

2. Gartner, "The Spam Within: Gartner Says One-Third of Business E-Mail Is 'Occupational Spam,'" April 2001, http://www.gartner.com/5_about/press_room/pr20010419b.html (accessed 3 Nov. 2005).

3. Simon Caulkin, "Adrift in a Parallel Universe," *The Observer*, 28 Aug. 2005, http://observer.guardian.co.uk/business/story/0,6903,1557681,00.html (accessed 3 Nov. 2005).

4. Tristan Leaver, personal interview.

5. Michael Pusateri, Elisabeth Freeman, and Eric Freeman, "Leveraging RSS at Disney: From Collaboration to Massive Content Delivery," O'Reilly Emerging Technology Conference, 10 Feb. 2004, http://conferences.oreillynet.com/presentations/et2004/disney.pdf (accessed 3 Nov. 2005). Also see Michael Pusateri, "Lightweight Apps at Disney," Web 2.0 Conference, 5 Oct. 2004.

6. Suw Charman, "Dark Blogs: Case Study 01: A European Pharmaceutical Group," 13 June 2005, http://www.suw.org.uk/files/Dark_Blogs_01_European_Pharma_Group.pdf (accessed 3 Nov. 2005).

CHAPTER SEVEN

Economic Blogs and Blog Economics

John Quiggin[1]

A s blogs have proliferated and diversified, it has become more difficult to discuss blogging as a unitary phenomenon, and more necessary to focus on particular categories and sub-categories of blogs; this, of course, is the impetus for this book. As recently as 2004, a fairly complete list of academic blogs could be maintained in the sidebar of the *Crooked Timber* group blog. Today, growth in numbers has made this impossible, and it is now necessary to look at particular disciplines of blogging.

With the arguable exception of law, economics is the academic discipline where blogging has been embraced most enthusiastically. There are various reasons for this, including the fact that day-to-day news and events provide opportunities for economic analysis, and the centrality of economics to a wide range of political issues.

Economic Blogging

The standard view of issues-related blogging is one in which blogs are primarily devoted to comment and analysis of news derived, directly or indirectly, from reports in traditional media. In negative presentations, particularly from defenders of old-style journalism, blogs are often presented as

> **Issues-Related Blogs:**
> blogs devoted primarily to comment on and analysis of public policy and social issues.

"parasitic" on the news-gathering efforts of the established media; on this, also see chapters two and three in this collection.

Whatever its validity in general, this picture is largely inapplicable to economic blogging. The "raw material" of economic news is derived mainly from the output of (mainly public) statistical agencies; reports by governments, industry groups, and non-government agencies; and, occasionally, academic journal articles and working papers. This material is commonly published along with press releases issued to news media.

A typical economic news story consists of a summary of the relevant press release, with comments obtained from one or more sources. The most common sources for such comments are economists employed by banks and other financial institutions, who stand to gain publicity for their employers in return for being readily available and willing to provide immediate comment. By contrast, academic economists tend to be harder to contact and more cautious in commenting.

Bloggers can and do comment on such reports. In most cases, however, the original sources are published on the Internet and may be reached directly using hyperlinks. In particular, government statistical agencies routinely publish information on the Internet. In most cases, a press release announcing new data is accompanied by a more detailed statistical report. This opens up a range of possibilities.

First, blogs provide space for more detailed analysis of statistical releases than is feasible in a newspaper report (let alone broadcast media). In addition, specialist blogs can assume a substantially higher level of awareness of relevant background than newspapers, again allowing for a deeper analysis.

Second, there is no particular limit to the space that can be allocated to a particular topic, and no need to give a full summary of the official release. If some particular aspect of monthly data is of special interest, it can be covered in detail.

Third, the capacity of blog posts to link both to original sources and to previous posts makes it possible to achieve continuous coverage of a topic without the repetition inherent in monthly reports.

Finally, newspapers are generally reluctant to report on academic working papers and similar publications unless the conclusions are obviously newsworthy. To the extent that such contributions are taken into account at all, it is through repetition of their arguments by quotable officials, or if they become part of the conventional wisdom of the day. By contrast, bloggers can and do report on new ideas being advanced in this way.

Econobloggers and Analysis

As has been mentioned already, economic bloggers are unusual in their capacity to compete with the news-reporting function of mainstream media. Nevertheless, as with other issues-based bloggers, the main interest is typically in the presentation of opinion and analysis.

In the United States, blogs already present serious competition for the mainstream media in their coverage of economic issues. In fact, someone seeking to be informed on current issues relating to economic policy would do better to read blogs than to follow the coverage in the mainstream media.

This assessment reflects both the strength of economic blogging in the United States, where dozens of academic economists (including Nobel Prize winner Gary Becker) offer blog comment on a range of issues, and the weakness of newspaper coverage of the same issues.

In contrast to both the explosion of the blog scene and the practice of the past (when both Paul Samuelson and Milton Friedman regularly wrote for the mass media), and with the notable exception of Paul Krugman at *The New York Times*, leading economists rarely contribute to mainstream media these days. Although economic analyses are occasionally published in the business sections of leading newspapers, they are fairly limited and too infrequent to constitute a serious basis for public debate.

In the United States, blogs fill an important void in mainstream media reporting on economic issues by presenting more studied opinions from both sides of the political spectrum. In a country where journalism is dominated by conservative political interests, "supply-side" theories of writers like Jude Wanniski and Arthur Laffer, regarded by the bloggers and most serious economists as "voodoo economics," are nonetheless the most common feature of economic reporting.[2]

The relationship between economic blogs and opinion columns is quite different in Australia. The mainstream media coverage of economic issues in Australia is superior to that in the United States, particularly in relation to opinion and analysis. The major Australian dailies all have long-standing and well-regarded economic writers (Alan Mitchell at the *Australian Financial Review*, Ross Gittins at the *Sydney Morning Herald*, Alan Wood at the *Australian*, Ken Davidson and Tim Colebatch at the *Age*) whose work regularly appears in the main opinion pages. This is in addition to comment and analysis in the business sections. In addition, the *Australian Financial Review* makes a more substantial contribution to debate than does *The Wall Street Journal*.

By contrast, the number of bloggers writing on economic issues in Australia is smaller than in the United States, although it is probably comparable relative to population. In addition, several of the bloggers who write regularly on economic topics (Nicholas Gruen, Andrew Leigh, Andrew Norton, and myself) are regular or freelance opinion columnists for major newspapers.

One result is that Australian economic blogging does not display the adversarial relationship between blogs and traditional media that is evident in many other contexts. For bloggers who work both sides of the street, blog posts can be used as sounding boards for ideas that will be worked up into opinion pieces, or as amplification of pieces that have already been published in the constrained space (of typically 700 words or so) of an opinion column.

Of course bloggers also link to opinion pieces in the mainstream media, and the nature of both media is such that conflict is more interesting than consensus, so links are more commonly critical than favorable. However, by comparison with the U.S. scene, the lines of debate are relatively fluid, and people who are generally on opposing sides may line up together on particular issues.

Blogging on Economic Policy Issues

Compared to newspapers and to traditional academic outlets such as journal articles and working papers, blogs have substantial advantages in discussing economic policy issues. First, although posts are typically aimed at the educated public rather than career economists, economic bloggers can assume a higher level of understanding than can journalists, even when writing for *The Wall Street Journal* or the business pages of *The New York Times*.

Second, the hyperlinking capacity of blogs makes cross-reference easier than for either newspapers or academic articles. A standard academic citation requires the reader to do the work of locating the cited article. This is changing as working papers are presented on the Web and with hyperlinks, but blogs are still well ahead.

Finally, blogs allow for a continuous discussion through a series of posts in a way that is not feasible either in newspapers, where it is difficult to sustain a continued discussion for more than a few weeks, or in academic journals, where the typical lead time for publication is more than a year.

The strengths of blog-based discussion may be illustrated in relation to debates over the sustainability of current account deficits.

A Blogging Case Study: Current Account Deficits

The sustainability or otherwise of the U.S. current account deficit is a central issue in economic policy debate. Analysis of the monthly trade statistics and quarterly current account statistics is therefore a matter of vital interest. Perhaps the most important distinction is that between the *trade balance* and the *current account balance*, which is the sum of the trade balance and net payments of interest and dividends to foreign owners of capital.

> **Trade and Current Account Balances:** the balance of payments on trade in goods and services (the trade balance) is the difference between the value of exports of goods and services and the value of imports. The balance of payments on current account (the current account balance) is the sum of the trade balance and net payments of interest and dividends to foreign owners of capital.

Despite the centrality of this distinction, *The New York Times* and even *The Wall Street Journal* routinely gloss the term "current account deficit," explaining that this is "the broadest measure of the nation's trade deficit." Implicit in such a gloss is the assumption that readers are encountering the term for the first time and cannot be expected to be aware of the way in which trade is treated in the national accounts, let alone of recent movements in the deficit and its various components.

A number of blogs pay close attention to current accounts and to the debate over whether the large current account deficits now being recorded in Australia and the United States are sustainable. Examples include

- *Brad Delong's Semi Daily Journal*, http://delong.typepad.com/;
- *Brad Setsers Web Log*, http://www.rgemonitor.com/blog/setser/;
- *Calculated Risk*, http://calculatedrisk.blogspot.com/;
- *Institutional Economics*, http://www.institutional-economics.com/; and
- my own blog, *John Quiggin*, http://www.johnquiggin.com/.

Two features of the blog debate over the current account deficit are noteworthy. First, bloggers who have commented on this issue generally take a sharply defined position, rather than covering the issue in the "balanced" fashion that is favored in mainstream news coverage and influences what is considered appropriate in more analytical pieces. (Krugman, a notable exception to standard media practice, is regularly criticized for being shrill.)

Second, the majority of bloggers who have written on the issue take the pessimistic view that deficits are unsustainable and that the process of adjustment is likely to involve a "hard landing." However, the opposite view—that current account deficits are simply an aggregate of voluntary transactions between "consenting adults" and can therefore never be a matter of policy concern—is also well represented, notably by Stephen Kirchner of *Institutional Economics*.

Overall, the blog debate is conducted at a much higher level of sophistication than that in the media. Arguments that the present arrangements are sustainable since they arise from either a "savings glut" in Asia or from an implicit international agreement along the lines of the old Bretton Woods Agreement are subject to sustained critique or defense, without the need to restate basic issues.

> **Bretton Woods Agreement:** the Allied Powers, meeting at Bretton Woods, New Hampshire, in 1944, established a system of fixed exchange rates that prevailed from 1945 to 1970. Some economists see a Bretton Woods II system emerging, based on stable exchange rates between the U.S. dollar and Asian currencies.

An important example of the greater sophistication of the blog-based debate relates to the distinction between the trade balance and the current account balance noted above. There is considerable debate about the sustainability of large current account deficits (above 5 percent of GDP). By contrast, it is a matter of basic arithmetic that consistent large deficits in the trade balance are not feasible, since the resulting accumulation of debt would lead to explosive growth in interest payments and the current account deficit.

It follows that, both in Australia and the United States, the trade deficit must be eliminated or greatly reduced over the next decade or so. Serious debate among economic bloggers turns on the question of whether this adjustment is likely to take the form of a painful economic contraction or whether it will occur smoothly as a result of strong growth in productivity and output.

These points are barely recognized in the mainstream media debate. Rather, it is widely assumed that, provided foreign investors are willing to supply the requisite flows of capital, both the trade and current account deficits can be sustained indefinitely.

The superiority of blog analysis is also evident at the level of detail. Blogs provide detailed coverage of issues such as the composition of trade and capital flows and the maturity profile of public debt that are essential to a complete understanding of the issues but too complex and dense for even the business pages of the mainstream media. The capacity to provide direct hyperlinks to official data is crucial here. Bloggers perform an important service of connecting isolated data from global sources in a manner that permits informed decisions to be made. This is a service that goes well beyond the local level: the impact of these economic decisions is global.

The value of blogs in the context of economic policy, then, is in addressing regional economic policy questions from a broader perspective, spatially and temporally. Because economists tend to read each other's work, global coverage is achieved; because the blog posts are archived and cross-linked, the longer-term ramifications become more visible. In the United States, someone wishing to be well informed on economic issues would do better to read a good selection of blogs than to follow the business pages of the major newspapers. While this may not always be true internationally, it is still the case that the most comprehensive coverage of economic issues is in the blogs. Consequently, a significant use of blogs in economics is to negotiate economic dispute. The sphere of economic blogging therefore provides a deliberative space for engagement with economic policy issues that is not unlike the model of deliberative journalism Bruns identifies in chapter two of this book.

Blog Economics

Economic blogging encapsulates the self-reflexive nature of blogging. In commenting on various aspects of the economy, bloggers are themselves engaged in a new kind of economic activity, as I have noted in the previous section, and one that has yet to be fitted into existing categories of economic analysis.

Considered as a component of the market economy, however, blogging is as yet of minimal significance. While there are a few celebrated cases of bloggers currently earning a full-time living through their craft, certainly there are no more than a handful doing so. Blog software companies including Blogger (now a subsidiary of Google) and Six Apart employ a modest number of programmers and support staff, but as yet the industry is small and immature. Blogs also require general Internet support services (hosting, ISPs) and so on, but the still dominantly text-based nature of blogging means that these demands are modest relative to, say, the peer-to-peer sharing of video and audio files.

Blogs can generate revenue from advertising, from the provision of referral services to online merchants such as Amazon, and from donations (tip jars). Of these, only advertising is likely to be significant in the long run. However, it appears at this stage to be a very modest source of revenue. The most prominent specialist provider of advertising aimed at blog audiences is BlogAds, which serves several leading bloggers including *Daily Kos*, *Wonkette*, and *Andrew Sullivan*. Of these, *Daily Kos* is shown as generating more than US$6,000 a week in ad revenue, but no other site generates more than $1,500 a week, which is probably the minimum required to generate a single full-time income after allowing for commissions and hosting costs.

However, economic analysis is not confined to activities that produce goods and services traded in markets. The most important components of non-market economic activity are services provided by governments and those produced and consumed within households. A third category is defined by formal and informal social groups such as clubs and churches.

Blogging combines elements of all three forms of activity. It is undertaken primarily by individuals seeking self-expression as writers and entertainment or information as readers, and in these respects it is most naturally considered as part of the household sector. The resemblance to clubs is also evident, with blogrolls providing statements of affiliation to some of the multitudes of overlapping communities that make up the blogosphere. Finally, and perhaps most importantly, the output of blogging is one of the most perfect examples of a pure *public good*, being both non-rival (apart from bandwidth costs) and not subject to exclusion. (This is technically feasible, but blogs that require any form of registration rarely attract readers.)

Public Good: as defined by economists, a public good is one that is both nonrival and nonexcludable. Nonrivalry means that one person's use of the good does not affect the amount available for others. Nonexcludability means that once the good is produced, no-one can be prevented from using it.

The "public good" nature of blogging means that non-monetary motives generally dominate. Some of the most obvious are new and blog-specific, notably including the economy of esteem embodied in the exchange of hyperlinks between posts on different sites, and the more permanent recognition embodied in blogrolls.

Although non-monetary motives for blogging are diverse, they tend to reinforce each other, or at least tend not to contradict each other. By contrast, market rationality tends to crowd out non-monetary motives. This point has been much discussed in relation to altruism,[3] but it arises in relation to other motives, too.

Consider self-expression, one of the major motives for blogging. On the one hand, the presence of commercial motives tends to constrain self-expression, leading to a self-censorship of any form of expression that might alienate advertisers or audiences. Even more seriously, to the extent that self-expression is a marketable commodity, it will be profitable to simulate it using market-tested standard formats (popular music provides many examples). This in turn raises a whole range of problematic issues about authenticity and commercialism.

The point is even clearer when we consider the more formalized interactions of linking and blogrolling. These interactions form the basis of an economy of esteem, with formal rankings provided by sites like *Technorati*, one of the main enterprises tracking blogs. A favorable (or even unfavorable) link from a high-status blogger provides a new or obscure blogger with the chance to present their work to a large group of visitors, some of whom may become regular readers.

Since links are desirable, they could be the subject of various kinds of trade, and there have, in fact, been occasional instances of sponsorship schemes that allowed for the possibility of purchasing places or blogroll links. Clearly, to the extent that such trade became widespread, the economy of esteem would be subverted. A positive link would be of no more significance, in terms of esteem, than a celebrity endorsement of a commercial product (though, as this example shows, it might nonetheless be commercially valuable).

These examples could be multiplied. Market rationality has a totalizing effect that tends to exclude other motives. It can be managed only by demarcating spheres in which the rules of market rationality do or do not apply. (Consider the frequent admonitions to keep business separate from friendship or family.)

The Economic Significance of Blogs

Having observed the difficulty of fitting blogging into traditional economic categories, it is worth considering some broad estimates of the economic significance of blogs, in terms of the economic resources used in their production, and the economic value of blogs as consumption items. Even in the absence of market exchange, these issues can be examined in terms of the traditional economic distinction between supply and demand.

On the supply side, the number of blogs is growing rapidly and is estimated by *Technorati* at over 15 million. However, because of the rate at which blogs are created and abandoned, and the fact that leading community sites automatically create blogs for new members, this figure is of doubtful significance.

The most important measure of the scale of blogging as an economic activity is that of labor input. *Technorati* reports that, at the end of July 2005, there were about 900,000 posts created each day. The number of posts recorded by *Technorati* peaks (relative to the underlying upward trend) on days of major news events, suggesting that issue-based blogging contributes a substantial proportion of posts.

Assuming that the average blog post takes an hour to prepare (this includes overheads like research, if any, site maintenance, responses to comments, and so on), 900,000 posts per day amounts to over 300 million hours per year, equivalent to the work of around 150,000 full-time, full-year workers. As a comparison, the U.S. Bureau of Labor Statistics reports that news analysts, reporters, and correspondents in the United States held about 66,000 jobs in 2002.[4]

Turning to the demand side, a survey conducted by the Pew Center early in 2005 indicated that 27 percent of Internet-using Americans had read blogs, 9 percent had created a blog (this is broadly consistent with the *Technorati* estimate cited above), but that 62 percent did not know what a blog was.[5] For comparison, Pew estimates that the number of adult readers of blogs is about 40 percent of the size of the talk radio audience and about 20 percent of the size of the newspaper-reading population. However, it may be that relatively few of the respondents are regular blog readers. Another study suggested that only 5 percent read blogs on a weekly basis.[6]

One way of reading this evidence, consistent with widely held preconceptions about blogging, is that the average blog has about three readers. This is misleading, since (as noted above) the number of blogs actually extant is a fraction of the number originally created, and most users sufficiently aware to distinguish blogs from other sites probably visit more than one blog. Still, when expressing the significance of blogs using the standard metrics of audience size that are applied to traditional media, significance must be deemed low, as the ratio of readers to writers is low.

But arguably an even more significant economic output of blogs and similar user-driven Internet activity arises from their role as drivers of technical innovation. Increasingly, innovations like RSS and trackback are flowing from blogs to the commercial sector rather than *vice versa*. Therefore, the significance of the technology, the impact, and the contribution of blogging to labor productivity may be regarded as substantially varying from the standard metrics of audience measurement that are usually applied to media products.

Measuring the Blog Economy

Blogging is, or at least has been so far, a quintessentially non-economic activity. Yet blogs are already significant in economic terms, and economic bloggers are making an increasingly important contribution to public debate. The question that arises, therefore, is: What measures ought to be used to determine the economic impact of blogs? Clearly, audience size is insufficient as a sole measure. Paid labor calculations are also inexact, as the voluntary labor market for blogs, as well as the industrial ramifications of the technology and the industry of blogging, are complex and as yet not measured. Further, the dominance of non-monetary benefits arising from blog links and blogging generally means that the economics of esteem are significant and ought to be considered in any economic measurement of impact. Given the public good aspects of blogging and the latent economic impact of blogs on the development of economic policy (as well as the corresponding impact of blogs in other fields on their specific areas of interest), metrics for economic significance of blogs need to go well beyond the net present value of the industry of blogging or the act of engagement with blogging. And considering the as yet relatively low penetration of blogging awareness among the total Internet population (as indicated in the Pew research cited above), if the growth rate in awareness and reading of blogs among Internet users continues to rise, then the significance of blogs may well compound. Finally, as blogs grow in number, and as blog audiences swell (even if not in ratio of readers to writers, then in blog penetration rates), that significance can only rise.

In combination, all of these observations point to a clear need for new approaches to measuring the economic impacts of blogging and other forms of large-scale, distributed, *ad hoc*, and only loosely organized, user-led forms of intellectual labor, creative practice, and technological innovation. Analyses that employ public good models and, like the work of Lawrence Lessig,[7] operate using the model of an electronically mediated network commons, may provide first glimpses of the emergent economic models in this field. The implications of this for economic organization are as yet unclear, but are likely to be profound.

NOTES

1. I thank Nancy Wallace for helpful comments and criticism. This research was supported by an Australian Research Council Federation Fellowship.

2. The term was coined by George Bush Sr. during the Republican primary campaign for the 1980 presidential election, won by Ronald Reagan. The standard professional assessment is given by Greg Mankiw, who later became head of the Council of Economic Advisers under George W. Bush. Mankiw's *Principles of Macroeconomics* textbook, 1st ed. (New York: Harcourt, 1997), pp. 29-30, refers to Ronald Reagan's supply-side advisers as "charlatans and cranks."

3. R. Titmuss, *The Gift Relationship* (London: Allen & Unwin, 1970).

4. Bureau of Labor Statistics, U.S. Department of Labor, "Occupational Outlook Handbook, 2004-5 Edition, News Analysts, Reporters, and Correspondents," 2005, http://www.bls.gov/oco/ocos088.htm (accessed 20 Sep. 2005).

5. Pew Internet & American Life Project, "The State of Blogging," 2005, http://www.pewinternet.org/press_release.asp?r=104 (accessed 20 Sep. 2005).

6. C. Li, "Blogging: Bubble or Big Deal? When and How Businesses Should Use Blogs," 2004, http://www.forrester.com/Research/Document/Excerpt/0,7211,35000,00.html (accessed 20 Sep. 2005).

7. Lawrence Lessig is the driving force behind the Creative Commons movement and its more recent offshoot, the Science Commons—see http:// creativecommons.org/, and http:// sciencecommons.org/ respectively. Also see Lessig's books on this topic: *Code and Other Laws of Cyberspace* (New York: Basic Books, 1999); *The Future of Ideas: The Fate of the Commons in a Connected World* (New York: Random House, 2001); *Free Culture: How Big Media Uses Technology and the Law to Lock Down Culture and Control Creativity* (New York: Penguin Press, 2004).

CHAPTER EIGHT

Blogging the Legal Commons

Ian Oi

There are certain natural affinities between legal writing and blogging. Legal reasoning in the common law tradition is steeped in the methods of judicial precedent as a way of thinking, a primary means by which coherence and legitimacy are lent to what might otherwise be attacked as incoherent or illegitimate. The weight of the "authority" of judicial decisions bears heavily on lawyers of the common law persuasion. Much of the young lawyer's early training is spent learning the techniques by which cases and judgments are to be compared, followed, distinguished, and above all *cited* in the course of documenting legal advice, legal advocacy, and (at the apotheosis) legal judgment. See for

> **Common Law:**
> the unwritten, judge-made law in countries such as England, Australia, New Zealand, Canada, and the United States of America, which has evolved under a centuries-old legal tradition of precedence, interpretation, expansion, and modification. Contrasted with **statutory law** made by legislatures, and the legal tradition of codified laws in Western European civil law countries such as France, Italy, and Germany.

yourself: pick up the written decision of any superior court judge in, say, a contract or tort case, and count the citations of past decisions of other judges, see the way counsel's arguments for the appellant or the respondent are summarized and then destroyed or adopted, how the judge justifies a disagreement with a fellow judge in the same case or a different case. These are complex threads of imagined conversations between lawyer and lawyer, woven fundamentally into the fabric of judge-made law.

The similarities between legal writing and the practices of lawyers as described above, on the one hand, and blogs and the practices of bloggers, on the other, make it unsurprising that a growing crop of lawyers has embraced blogging. As we will see from a short survey below, these lawyer-bloggers range from legal academics to private practitioners to judges.

In pondering these questions, we incidentally notice another convergence at work. At the most superficial level, blogging gives lawyers another way of talking *at* readers. Those readers are sometimes actually other lawyers, but the diarizing, conversational language of many lawyer blogs—in common with other blogs—leads one to suppose that the imagined reader is mostly the educated layperson. Yet the nature of the blog is such that it looks beyond reader passivity and invites reader participation. Where the reader is another lawyer, the invitation can be inferred as one of participation in legal discourse along (relatively) familiar lines of critique and common law making, as described above. Where the reader is the non-lawyer—as the imagined blog reader mostly is—the legal blog invites new forms of creative collaboration. Economics, politics, culture, and technology: these subjects are alien to "proper" legal discourse but are inevitably brought to the writing table by the lay reader in considering and commenting on the legal blog. Also inevitably, these contributions go beyond reactive, annotative comments on lawyer blogs; we start to see informed, thoughtful, and interesting legal blogging by non-lawyers, bouncing off lawyer blogs and in turn generating considered blogging responses from lawyer blogs.

This confluence and conference of lawyers and non-lawyers thus creates a web of legally focused content that is self-referential (like Law) yet understandable by and accessible to the newcomer (like Blog); topical (like Journalism) yet persistent (like Judgment), focused on certain concerns (the Legal) but open to different approaches to those concerns (the non-Legal). Tellingly, when U.S. Federal Circuit Judge Richard Posner and Nobel Prize-winning economist Gary Becker decided to start their joint blog *The Becker-Posner Blog*[1] in December 2004, they explained their decision in these terms:

> blogging is a major new social, political, and economic phenomenon. It is a fresh and striking exemplification of Friedrich Hayek's thesis that knowledge is widely distributed among people and that the challenge to society is to create mechanisms for pooling that knowledge. The powerful mechanism that was the focus of Hayek's work, as of economists generally, is the price system (the market). The newest mechanism is the "blogosphere." There are 4 million blogs. The internet enables the instantaneous pooling (and hence correction, refinement, and amplification) of the ideas and opinions, facts and images, reportage and scholarship, generated by bloggers.
> We have decided to start a blog that will explore current issues of economics, law, and policy in a dialogic format.[2]

> **Legal Commons:** a subset of the blogosphere populated by lawyers and interested laypersons, debating legal and related issues.

Such descriptions of the web of legal blogging content are an apt summary of an emerging commons of knowledge: a Legal Commons built on and in an uncommon medium for the development of a body of legal thought.

This chapter describes an Australian pocket of the Legal Commons, and explores how it manifests an issue that may (or, intriguingly, may not) be central to its structure: the legal foundation for the Legal Commons.

One Australian Legal Commons

One of the core concerns of the present generation of Australian legal bloggers is the collision of technology and law (particularly that relating to intellectual property). Thus, a prominent legal blog in Australia has been *Weatherall's Law: An Intellectual Property Blog from Oz*,[3] a personal blog established by intellectual property (IP) academic Kim Weatherall in June 2002. *Weatherall's Law* initially started by noting newsworthy developments in intellectual property, technology and cyberlaw relevant to Weatherall's academic interests,[4] annotated with pithy commentary. It has since evolved to include sustained analyses of topical IP issues over the course of several postings, critiques that are periodically consolidated and revised. Copyright-focused topics that have been treated in this way on *Weatherall's Law* include:

- the Australian litigation concerning the Kazaa peer-to-peer filesharing software, and related issues, often with overseas comparisons;
- the statutory and judicial treatment of technological protection measures law in Australia and overseas; and
- current legislative reviews of copyright law, for which her blog collects, summarizes, and reviews various submissions.

As to her reasons for blogging, Weatherall has this to say:

it seems like every two seconds someone in the blogosphere asks why people blog—in particular, why academics blog.

Personally, I think I see benefit from it. In terms of reputation, profile, and getting invited to conferences—as well as getting feedback on ideas and the like.

Does the benefit outweigh the time costs? Perhaps not for someone chained to their computer anyway. But for me, it's become an ordinary part of my life.[5]

Weatherall's Law often refers to other Australian legal bloggers—lawyers and non-lawyers—and is itself often referred to by other bloggers on legal issues, again by lawyers and non-lawyers.[6] Indeed, it is arguably at the hub of a legal blog commons. *Weatherall's Law* has since generated the group blog *Law Font*, in which Weatherall joins with a group of like-minded lawyers to blog "an analysis of law, technology, economics, and policy."[7] *Law Font*, which commenced in June 2005, states its intention of covering a broader range of topics than *Weatherall's Law*:

Law Font is a group blog dedicated to discussing issues with a tech and law flavour. Because of the nature of the subjects, and the bloggers, the discussion often flows into economics, policy, and the occasional bit of pure geekiness.

All of the contributors work in the law, with some element of tech mixed in. The blog reflects our multi-jurisdictional interests, too.[8]

One of the lawyer blogs sometimes blogged by *Weatherall's Law* is the personal blog of Melbourne intellectual property barrister Warwick Rothnie, *IP Wars*,[9] which is similar in style and substance to the early postings of *Weatherall's Law*. Another is Queensland lawyer David Starkoff's personal blog *Inchoate*,[10] which focuses more closely on esoteric and eccentric aspects of Australian law and lawyers. Each of these Australian lawyer blogs also cites (and links to) *Weatherall's Law* on occasion, for a piece of news first reported there or a bit of analysis that has been blogged. These blog citations in turn elaborate, on occasion, on the source citation.[11]

Non-lawyer Australian blogs that are often referred to by *Weatherall's Law* (and which, in turn, often refer to it) include the personal blog of academic economist John Quiggin,[12] styled as a "Commentary on Australian & World Events from a Social Democratic Perspective," and *Rusty's Bleeding Edge Page*,[13] the personal blog of Free Software programmer Rusty Russell. The blogs of Quiggin and Russell have engaged with Weatherall in constructive dialogue on a number of legal topics, offering non-lawyerly perspectives that Weatherall has in turn taken account of in further developing her thoughts on those topics within *Weatherall's Law*. An example of this dialogue—in the context of the licensing of the Legal Commons—is described below.

Weatherall's Law also refers to leading non-Australian legal blogs for significant material, drawing attention to their relevance for an Australian audience and comparing and contrasting their content with local developments. These blogs include:

- *lessig blog*,[14] the personal blog of U.S. legal academic Lawrence Lessig, a founder of the Creative Commons movement;
- the personal blog of Canadian legal academic Michael Geist,[15] and
- *The Becker-Posner Blog*.[16]

To License or Not?

Surprisingly, perhaps, as these blogs refer to one another's contents, the legal status of the material which is referred to, cited, commented upon, and reworked in new blog postings on another site is rather murky—indeed, in spite of the intellectual property focus of the blogs, the question of content licensing in this case mirrors similar questions in the rest of the blogosphere.

The content licensing terms expressed at each of those blogs can be summarized as follows:

Weatherall's Law	U.S. Creative Commons License: Attribution—Noncommercial
Law Font	U.S. Creative Commons License: Attribution—Noncommercial
IP Wars	Copyright ownership notice with "all rights reserved"
Inchoate	Australian Creative Commons Licence: Attribution—Noncommercial—No Derivatives
John Quiggin	Australian Creative Commons Licence: Attribution—Noncommercial—ShareAlike
Rusty's Bleeding Edge Page	Australian Creative Commons Licence: Attribution
lessig blog	U.S. Creative Commons License: Attribution
Michael Geist	Canadian Creative Commons Licence: Attribution
The Becker-Posner Blog	U.S. Creative Commons License: Attribution—Noncommercial

Table 8.1: Content licenses in selected Legal Commons blogs

As is apparent from the above, all of the blogs bar one use licensing provisions that explicitly control the uses that might be made of their blog content. The one exception—*IP Wars*—does not, however, appear to discriminate regarding the use it makes of the other blogs' content,[17] or to be discriminated against in the use made of its content in other blogs.[18]

The licensing employed by the other blogs described above is the Creative Commons scheme,[19] which seeks to use explicit "some rights reserved" licensing as a means of building "a reasonable layer of copyright."[20] Salient features of Creative Commons licensing for present purposes are that:

- There is no single Creative Commons license. Rather, there is a suite of Creative Commons licenses, each comprising a different combination of a limited set of licensing attributes.
- The original Creative Commons licenses were based on U.S. law. Subsequent "translations" of the licenses for the laws of other jurisdictions have been made.[21] Note that each Creative Commons license—regardless of the jurisdiction for which it is tailored—is expressed as a worldwide license of the rights licensed by it.

> **Creative Commons:**
>
> a non-profit organization that offers legal document, technological, and social network tools that enable authors and artists to license their material with a flexible range of protections and freedoms. The suite of Creative Commons licenses builds upon the "all rights reserved" of traditional copyright to create voluntary "some rights reserved" copyright regimes.

Nonetheless, the fact that the other blogs do employ a licensing scheme that explicitly sets out use permissions and restrictions should not be taken as an unambiguous sign that those other bloggers accept the necessity of licensing as a means of controlling use of their material by others. This hesitation is most clearly expressed by Weatherall, in explaining her adoption of the Creative Commons licensing scheme:

> Creative Commons licenses strike me as a *peculiarly American solution* to a problem that we don't necessarily have in spades here (yet). The attitude of "Freedom and Contract for all" is terribly American. Yes, there is an international "porting" movement, which we have seen in Australia—but what I'm saying is that the *idea* of this kind of license is a bit American, and we should recognise the differences between the US system and ours.
>
> I would far rather see our fair dealing defences broadened a (sensible) bit, ensuring that we all have certain, sensible freedoms to build on what has gone before, than I would see Creative Commons licensing take off. If, *and only if*, such sensible reworking of fair dealing defences is a pipe dream—as it may be in the US, but isn't necessarily in Australia given the current review—*then* we should focus more energy on building that sensible layer.
>
> All that said, why do I apply a Creative Commons license to my material? Why buy in to this American solution? Because it seems to give people some comfort to see my positive affirmation on the blog that people may re-use the material. In other words, I'm using it not as a legal tool, but as an indicator of attitude. I suspect I'm not the only one ... .[22]

Thus, Weatherall blogs a first preference for an Australian legislative solution that recognizes the legal requirements of blogging and sees only symbolic, "last resort" use for Creative Commons in the Australian blogosphere.

John Quiggin responds to Weatherall by blogging a substantive reason for his employment of Creative Commons licensing. For economist Quiggin, it provides (in the particular configuration of Creative Commons licensing attributes he has selected) the best "default rule" for his kind of blogging:

> I've chosen the non-commercial, attribution, share-alike version of the Creative Commons License. This says that anyone can reproduce my work from the blog, with attribution and for non-commercial purposes, as long as they share it under the same conditions.
>
> I've chosen this, not because it's necessarily the best option in all, or even most cases, but because it's the best default rule ... . If someone wants to do things differently they can propose a contract with me. The optimal default rule is one that protects most rights I might want to enforce, while allowing (without special permission) most uses I'd be unlikely to object to. Public domain fails on the first count, and standard copyright on the second. I think the Creative Commons License, in the particular form I've chosen, gets the balance just about right.
>
> The general idea of a default value is familiar to anyone who's done any computer programming ... .[23]

By comparison, Rusty Russell (who selects the least restrictive of the Creative Commons licenses, requiring only attribution by a user of his blog material) blogs his disagreement, from the perspective of a Free Software programmer, with the significance accorded by Quiggin to his "default rule" analysis:

please note that blog licensing is much less important than software licensing to me, since (1) most uses of my blog will be excepted from copyright anyway under fair use/dealing/reasonableness, and (2) a handful of people (hello to you both) read my blog, millions are touched by my code.[24]

Weatherall duly blogs this debate,[25] chipping in her support for Russell's position as against Quiggin's "default rule" justification for Creative Commons licensing:

ah, you say—but you license your blog under creative commons. Well, Yes, I have a creative commons license on the blog. The reason? Because it sends a *normative* message—along these lines:

"feel free to copy this stuff and pass it on. And oh, if you can mention where you got it, that would be *really cool* because I'm an academic and basically a self-promoter."

And I'm just about happy with that message.

Rusty is right on this one. This matters *far less* in this context than it does for software licensing. My view at the moment is that the most important things to worry about in this space are indeed,

- code licensing (because code is functional and modular and you need lots of little bits together working on the same terms to create the right "ecology"); and
- open access to research.

But maybe I'll be proved wrong in the end. Will be interesting to see.[26]

The above summary of the "Great Blog Licensing Debate"[27] cannot do justice to the complexity of the debate that was conducted in the Legal Commons over the issues of: whether blogs should be licensed; if so, the justifications for such licensing; and the terms of such licensing. As well as the blog entries described above, incidental comments posted by passing blog readers also inflected the debate and at times influenced the course of argument. It is only through reading and clicking through the participating blogs directly that one gains a full appreciation of this collaborative exploration by lawyer and non-lawyer of the implicated legal issues.

Some concluding observations may be made regarding matters of substance emerging from the debate. First, it is curious that notwithstanding Weatherall's objection to the unthinking adoption of American licensing philosophies in an Australian environment, *Weatherall's Law* (and its progeny *Law Font*) chose to adopt, without critical comment, the U.S. form of Creative Commons licensing rather than an Australian Creative Commons Licence adaptation. One may speculate that a reason for this may be that the "mere" symbolic value of Creative Commons licensing—if that is its primary value for a blogger—is best exemplified by the use of the U.S. form of license.

Second, it is interesting that a significant proportion of the legal bloggers described above were content to license their works under a "bare" Attribution license (i.e., almost No Rights Reserved, with no restrictions other than the requirement of attribution of the author). The contrast with the default customarily adopted by orthodox, non-blogging Websites—"All Rights Reserved" for all uses of the provided Web content—is striking.

Finally, the evident lack of uniformity in approach to blog licensing issues—particularly among a group of exceptionally well-informed legal bloggers, lawyers, and non-lawyers alike—suggests scope for further development of this particular debate within the Legal Commons. However, the fact that such lack of uniformity of approach has apparently not impeded the progress of that debate through the blogging apparatus gives practical perspective to these issues: maybe there really are more important things to blog about in the Legal Commons than the Law of the Blog. What this debate, and others like it, demonstrate clearly, at any rate, is that there *is* a strong sense of a Legal Commons among lawyer and non-lawyer bloggers in this region of the blogosphere.

Notes

1. See http://www.becker-posner-blog.com/.

2. Gary Becker and Richard Posner, "Introduction to the Becker-Posner Blog," *The Becker-Posner Blog*, 5 Dec. 2004, http://www.becker-posner-blog.com/archives/2004/12/introduction_to_1.html (accessed 16 Nov. 2005).

3. See http://weatherall.blogspot.com/.

4. Weatherall is currently associate director of the Intellectual Property Research Institute of Australia, and lecturer in law at Melbourne University.

5. Kim Weatherall, response to "Experiments in Scholarly Publishing," *Law Font* 24 (Oct. 2005), http://www.lawfont.com/2005/10/20/experiments-in-scholarly-publishing/ (accessed 16 Nov. 2005).

6. For instance, a *Technorati* (http://www.technorati.com/) search of *Weatherall's Law*, conducted on 14 Nov. 2005, indicated 151 "Newest Posts" links to the site on 62 Websites.

7. For the moment, posts on *Weatherall's Law* are also cross-posting to *Law Font*.

8. *Law Font*, "About *Law Font*," n.d., http://www.lawfont.com/about-lawfont/ (accessed 16 Nov. 2005).

9. See http://homepage.mac.com/wrothnie/iblog/.

10. See http://www.dbs.id.au/blog/.

11. For example, the refusal of an application to the High Court of Australia for special leave to appeal in *The Panel* litigation was blogged on *Weatherall's Law* (http://weatherall.blogspot.com/2005_10_01_weatherall_archive.html#112866184336268311). This blog entry was itself then blogged on *Inchoate* with additional commentary and analysis (at http://www.dbs.id.au/blog/law/panels-end.html).

12. See http://johnquiggin.com/. Quiggin is also a contributor to this book—see his chapter seven on "Economic Blogs and Blog Economics."

13. See http://ozlabs.org/~rusty/index.cgi.

14. See http://www.lessig.org/blog/.

15. See http://www.michaelgeist.ca/.

16. See http://www.becker-posner-blog.com/.

17. See, for instance, *IP Wars* blog of *Weatherall's Law* coverage of *Stevens v. Sony* at http://homepage.mac.com/wrothnie/iblog/C688984015/E1053368670/index.html.

18. See, for instance, *Weatherall's Law* blog of *IP Wars* coverage of *Stevens v. Sony* at http://weatherall.blogspot.com/2005_10_01_weatherall_archive.html#112863924865713606.

19. See http://creativecommons.org/ (international) and http://creativecommons.org.au/ (Australian Website). For a full description of the different Creative Commons licenses, see http://creativecommons.org/about/licenses/meet-the-licenses.

20. *Creative Commons*, "'Some Rights Reserved': Building a Layer of Reasonable Copyright," n.d., http://creativecommons.org/about/history (accessed 16 Nov. 2005).

21. The author has been a co-project leader and a primary drafter of the Australian Creative Commons Licences.

22. Kim Weatherall, "The Great Blog Licensing Debate Continues...," *Weatherall's Law* 3 July 2005, http://weatherall.blogspot.com/2005_07_01_weatherall_archive.html #112038088651934888 (accessed 16 Nov. 2005).

23. John Quiggin, "The Creative Commons as a Default Rule," *JohnQuiggin.com* 20 July 2005, http://johnquiggin.com/index.php/archives/2005/07/20/the-creative-commons-as-a-default-rule/ (accessed 16 Nov. 2005).

24. Rusty Russell, "Licensing and the Knee-Jerk Reaction against Commercial Entities," *Rusty's Bleeding Edge Page*, 3 Aug. 2005, http://ozlabs.org/~rusty/index.cgi/IP/2005-08-03.html (accessed 16 Nov. 2005).

25. Kim Weatherall, "Blog Licensing. Yes, It Goes On," *Weatherall's Law*, 11 Aug. 2005, http://weatherall.blogspot.com/2005_08_01_weatherall_archive.html#112375273187482 142 (accessed 16 Nov. 2005). See also the blog entry references at http://weatherall. blogspot.com/2005_07_01_weatherall_archive.html#11205515710 3394007. Other blogging participants in this legal blog debate were David Starkoff (see http://www.dbs.id.au/blog/meta/blog-relicensing.html) and Adrian Sutton, a software programmer who blogged Starkoff's blog comments (at http://www.symphonious.net/2005/07/03/inchoate-relicensing/). Sutton's intervention provoked further blog commentary by Starkoff and blog responses from Weatherall.

26. Weatherall, "Blog Licensing. Yes, It Goes On."

27. So called by Weatherall: "The Great Blog Licensing Debate Continues...," *Weatherall's Law*, 3 July 2005, http://weatherall.blogspot.com/2005_07_01_weatherall_archive.html #112038088651934888 (accessed 16 Nov. 2005). Also called the "Australian Blog Licensing Frenzy" by Rusty Russell: "Creative Commons Licensing: Just Say Yes," *Rusty's Bleeding Edge Page*, 5 July 2005, http://ozlabs.org/~rusty/index.cgi/IP/2005-07-05.html (accessed 16 Nov. 2005).

Blogging to Basics:
How Blogs Are Bringing Online
Education Back from the Brink

James Farmer

The machine world has progressively killed off our authentic voice... We worked to get marks at school rather than to learn about our world and our place in it. We end up in jobs where we give up ourselves. We are bombarded with messages that tell us what we should look like, what we should wear, who we should mate with, how to be happy and who we should be. No wonder we find relationships difficult. How can we have a relationship with another when we have lost the core relationship with our own selves? —Robert Paterson (2005)

Robert Paterson seeks, in his compelling contribution to the 2005 "More Space" publication, to demonstrate how technology has impacted upon our lives. He asks us to consider how we relate to each other, our careers, our environment, and ourselves, and in doing so imagines a future in which blogs, open source, and emerging technologies have facilitated a modern-day classicism where we are "going home again to the place where humans fit."[1]

It is particularly significant that Paterson highlights our schooling and of equal interest that he should suggest that technology might have a holistic impact on the challenges that face education. While technology has been embraced throughout the school and higher education establishments and curricula, it is infrequently related to a rediscovery of more authentic learning and often argued to have had quite the opposite effect.[2] Consequently, to present an argument that more online engagement could help teachers and learners reconnect with themselves might be met with skepticism, yet this is exactly what this chapter aims to do. It presents the contention that online education is in a kind of pedagogical crisis, that an extensive reconsideration of approaches to teaching and learning online is necessary, and that the blogs and their associated technologies have the potential to develop into tools that can facilitate a transformation, should we choose to allow them.

Online Education in a Kind of Pedagogical Crisis

I believe that all education proceeds by the participation of the individual in the social consciousness of the race ... I believe that the only true education comes through the stimulation of the child's powers by the demands of the social situations in which he finds himself. (Dewey 1897, p. 77)

Only through communication can human life hold meaning. The teacher's thinking is authenticated only by the authenticity of the students' thinking. The teacher cannot think for her students, nor can she impose her thought on them. Authentic thinking, thinking that is concerned about reality, does not take place in ivory tower isolation, but only in communication. (Freire 1970)

Both John Dewey and Paulo Freire recognized and frequently highlighted in their work the critical importance of the social experience in education and the crippling effects of transmissive pedagogies and the systems that support them. Dewey argued that education that offers a pre-organized body of knowledge for transmission bred docility, receptivity, and obedience,[3] while Freire called for an end to the "banking" model of education which he saw as a process resulting in people being "filed away through the lack of creativity, transformation, and knowledge."[4] While their perspectives and contexts varied significantly—Dewey looking for enlightenment in Victorian England and Freire striving for freedom from the "ideology of oppression" in 1970s Brazil—these objections and their proposed solutions of social participation, through communication, have influenced generations of theorists and practitioners.

> **Online Education:**
> the increasingly popular use of Internet-based technologies to run on- and off-campus courses. Often supplemented by CD-ROMS and print materials.

However, in approaching online education it can be argued that we have, through ignorance or an overly technologically determinist leaning, found ourselves struggling with transmissive and communicatively-challenged pedagogies. In 2002 Morten Flate Paulsen as part of a study tour of Australia found that over 95 percent of Australian Universities surveyed were using one of two major Online Learning Environments (OLEs), and while it is reasonable to argue that there are significant differences between these particular environments,[5] it is equally possible to point to similarities. Both OLEs focus particularly on the management and distribution of content; both have been described at some point as "Management" systems; both have an enormously broad and powerful range of administrative tools; and, perhaps most significantly, both have one single prominent communication tool, the discussion board.

Of particular relevance in examining the pedagogical effectiveness of the discussion board is Randy Garrison and Terry Anderson's *E-Learning in the 21st Century: A Framework for Research and Practice.*[6] In this work the authors suggest that "a community of learners is an essential, core element of an educational experience when higher order learning is the desired learning outcome" (p. 22) and that "the idealized view of higher education, as a critical community of learners, is no longer just an ideal, but has become a practical necessity in the realization of relevant, meaningful and continuous learning" (p. 23). To achieve this, however, a prerequisite is effective communication, which is "at the heart of all forms of educational interaction" (p. 23) and which is critical to the formation of any community of inquiry. This is defined as being the product of effective social, cognitive, and teacher presence.

In a discussion board environment, it is difficult for users to develop effective social presence. For example, in a face-to-face context, individuals are able to project themselves in many ways, primarily through verbal and physical contributions to the people present in the area. However, in a discussion board, as well as being limited to the ability to express themselves through text, users are unable to express themselves to people in the area

> **Social Presence:**
> users are able to "project themselves socially and emotionally, as 'real' people." (Garrison & Anderson 2003)

because there may not be any people there. A contribution can be viewed and read by one person, the whole group, or nobody, and since how writers understand the intended audience of their work dramatically impacts on their entire approach to the task of writing,[7] this uncertainty impacts considerably on the ability of individuals to project themselves. Further, it is worthwhile to note the increasing use of detailed signatures on discussion board postings around the Web, as this is arguably due to the need, as seen by users, to project and convey themselves as real people (the signature may contain a picture, a link to a personal Website, a quote, or any other identifying characteristic) and in this sense demonstrates the inadequacies of the traditional discussion board model in the same ways that the emergence of emoticons has demonstrated the inadequacies of text-based email.

In a face-to-face context, a statement or question, particularly as part of a discussion or structured class environment, generally elicits a response from someone within that area. The utterance can be directed toward an individual or a group, and a following utterance can be expected. This forms the basis of any discourse in which meaning can be constructed. However, in a discussion board it is not possible to know who, if

> **Cognitive Presence:**
> users are able to reflect on their thoughts and "construct and confirm meaning." (Garrison & Anderson 2003)

anyone, will be reading an utterance, when this will occur, or, unless the user is permanently logged in to the discussion board and regularly hitting the refresh key, the moment at which this occurs. This is not dissimilar to entering a room that may or may not be frequented by the people you wish to communicate with (who will, in either case, be invisible to the user), leaving a message on the table, and then returning each day to see if someone has responded to the message. Likewise, any person responding to the message would have to visit the room each day to see if the writer or anyone else has replied to it. The room may be one of many rooms (there are frequently numerous discussion boards used in a single course), and there may be little or no reason other than to check for messages or responses that a person may choose to visit it. After several days of this kind of discussion, it is likely that in many cases a user will visit the room less, if at all.

In a discussion board environment, the teacher is rendered unable—through the technology—to exert any more influence than a student and hence faces considerable challenges in designing, facilitating, or directing "cognitive and social processes." As an experienced teacher and writer in the area of online teaching and learning remarked to the author, while discussing the use of discussion boards from a teacher's perspective, "they just bypass me and ignore me, it's like I'm not even there!"[8] When considered in a face-to-face context, this is not dissimilar to enforcing that teachers not stand, not position themselves any differently to the learners, and not use a whiteboard or any form of presentation. While this might be seen to be advantageous by some, its impact on facilitating the development of effective teacher presence is significant.

> **Teacher Presence:** a teacher is able to design, facilitating or directing "cognitive and social processes." (Garrison & Anderson 2003)

Consequently, online education has found itself in a place where, aside from content delivery, there is little opportunity for successful communication and much distance from the ideals of Dewey and Freire. Indeed, with the rising dominance of the established OLEs in the higher education marketplace and increasing exposure of on-campus as well as off-campus courses to the technology, there is an increasing danger that without a significant shift in the understanding of teaching and learning online, higher education will find itself a prisoner to the technology it has invested so much in.

Reconceptualizing Teaching and Learning Online

It is important to note, however, that while there is much to be desired in the dominance of particular products and modes of communication, many excellent teaching and learning experiences have taken place using these technologies. Discussion boards and commercial learning environments have been used to engage learners who previously would have had to make do with little more than a set reader and assessment guidelines, have allowed learners to complete studies in ways that would not previously have been possible, and have facilitated innumerable valuable conversations and exchanges of knowledge. Gilly Salmon has written widely about effective use of discussion boards in short course contexts,[9] and there has also been much discussion of the successful use of tools such as email[10] and wikis.[11]

What is particularly relevant in these cases is the means by which these successes have been aided through the technology used. Email, for example, is not particularly successful if one wishes to return to previous postings and is a very limited tool for the management of knowledge, wikis have little value in communication terms as in their purest forms they struggle to facilitate conversations, and the challenges posed by discussion boards have been well documented. An obvious conclusion to be drawn from this is that the technology chosen must be appropriate for the task undertaken. In this vein, it is pertinent to consider the necessary challenges faced by online educational practitioners and the appropriate uses of blogs where they might fit into a reconceptualization of what it means to teach and learn online.

I would propose, in particular, that this reconceptualization is best thought of in terms of the development of a community of inquiry and of Dewey's and Freire's insistence on the importance of communication and the role of the individual in the experience. As OLEs have evolved and email has come to be used generally as a tool of direct and unsustained communication (a noted researcher recently commented that the listservs he belongs to have become "little more than announcements and job postings"[12]), the environments that we communicate in have forced teachers and learners to focus on discussion boards and shared communicaton spaces, rather than on the individuals who are taking part in them. As a result of this shift, we have moved away from communication being centered on the individual and toward it being centered on an abstract space in which the individual participates. This is practically visible in the segmentation of learners into different courses, units, and "sections" that have little or no interrelation or integration.

Blogs allow us to challenge this direction, and this has been evidenced in a number of successful experiences using them in teaching and learning online. Gibson[13] found that his approach of providing each student with blogs and then exploring a number of alternative means of aggregation was particularly successful on a purely quantitative measure, with 31 students contributing a stunning 845 posts. Similarly, MacColl et al.,[14] while constrained to a degree in their use of aggregation (their learner blogs were placed behind an institutional firewall, and they do not comment on the impact this had on aggregation), focused their blogs on individuals and concluded that "overall we view the introduction of blogs into our studio courses as successful."

> Centered
> Communication:
> communication that
> is centered on the
> individual rather
> than the group.
> (Farmer 2005)

While it is not my intention to suggest that we should reject all forms of centralized communication, it is my contention that blogs provide what I have come to call "centered" communication[15] and as such are able to re-introduce the individual to the online learning experience. This is critical, I believe, in considering the development of a community of inquiry.

In terms of establishing social presence, it can be argued that blogs offer a significant opportunity for users to project themselves as "real" people. Bloggers are primarily writing to their own area and context, designed to their liking (if the blogger is not a Web designer, there is a wide range of templates available with every provider), and expanding on their previous postings from the online persona they have developed. Indeed, the fact that bloggers are also able to retain ownership of their writing, edit at will, refer to previous items and ideas, and control in its entirety the space and manner in which the blog is published, can significantly augment their control over their expression and hence increase the opportunity to project and the motivation for doing so.

Further, while the primary tool of communication in Weblogs is text, users are equally able to "photoblog," "audioblog," or "videoblog" their entries, and all of these kinds of projections are made to an audience that the blogger may well be largely aware of through currently evolving tools indicating subscribers to Webfeeds as well as their hyperlinked "blogosphere." Hence bloggers are able to express themselves through multiple media and assess, at the least, their immediate audience and, to an extent, their wider readership.

However, simply to be able to project oneself in online communication is not to be able to necessarily "construct and confirm meaning through sustained reflection and discourse in a critical community of inquiry," as stipulated by Garrison and Anderson as necessary for the development of effective cognitive presence.[16] Blogs undoubtedly support sustained discourse, as evidenced by the development and spread of memes and the ever-developing nature of the blogosphere,[17] but a question asked by many engaging with the technology is the extent to which this discourse is reflective, critical, and purposeful. While, for example, this kind of discourse is not apparent in the majority of blogging systems developed largely for socializing among teenagers, the charge that this invalidates the medium is inappropriate because of the evident breadth of uses of blogging.

A blog is a reflective medium (hence the comparisons with and the use of blogs as journals and diaries), and the nature of publishing to an audience in a manner that will be archived, can be referred to, and for which the author maintains responsibility and ownership, has developed a certain style of expression. Certain research[18] across the blogging spectrum has indicated that there is a possibility that Weblogs encourage significantly more in-depth and extended writing than does communication by email or through discussion board environments, though not as much as more formal modes of publication. The result, in an academic sense, is a kind of discourse somewhere between the conversational and the article. The value of this is evidenced through numerous examples of academic bloggers taking advantage of blogs in order to engage with their peers and students and to reflect on their own learning.

It could be argued that in terms of facilitating effective teacher presence, blogs are even less potent than discussion boards in their ability to empower the teacher to design, facilitate, or direct cognitive and social processes toward valuable educational outcomes. This argument can be based on the premise that a teacher's blog is essentially entirely separate from a learner's, and that the learner is under no obligation to read the teacher's blog. However, Clay Shirky has observed in his essay "Power Laws, Weblogs and Inequality"[19] that invariably, Weblogs fall into a "balance of inequality" in the same way that any system does if allowed "the very act of choosing" which, "spread widely enough and freely enough, creates a power law distribution." Effective use of blogs by teachers arguably places them as an organic central node to the class, and given the simplicity with which students would be able to aggregate their Webfeed and the selective "push" nature of this kind of aggregation—where Webfeeds are, despite being "pulled" by the user's aggregator, apparent to the end user as a pushed form of communication in much the same way as email—it is far more likely that the teacher will be able to facilitate and direct cognitive and social processes.

In terms of design, however, it appears in a basic sense that there is little in blogs that can be controlled in order to reach this outcome. While discussion boards can be placed alongside content in packaged courses and with limited opportunities to use the technology in ways unforeseen by the designer, a blog is essentially free-form, and there is little—besides providing templates, guidelines, and facilitating the group as a whole—that the teacher can do to actively impact on the technical structure of their experience. However, to introduce the metaphor of the traditional classroom, it is reasonable to ask to what extent cognitive and social processes can be impacted upon, and to what degree this is desirable within any context. For example, if a class is rigidly structured so that activities take place at fixed and inflexible times, if it contains set subject matter, and if students are assessed through standard matrices, the teacher takes little consideration of the benefits of learner-driven experiences and the vastly different requirements and approaches of individual students, as stressed by Gardner and Hatch.[20]

Naturally, this is not to say that an anarchistic structure is appropriate, but rather to suggest that one of the key attributes of Weblogs is that they have within them an "incorporated subversion"[21] that allows learners to express themselves and explore their context in ways independent of the original designers' intentions:

> Rather than design with constraint in mind, design with freedom and flexibility in mind. ... This emphasizes the active and purposeful role of learners in configuring learning environments to resonate with their own needs, echoing the notions of learning with technology through "mindful engagement."[22]

Blogging to the Future

A common charge levelled at the use of blogs is "why would I want one?" and this is entirely understandable, given their still-predominant use as tools for the publication of opinion, lifestyle, and—to a lesser degree—news. However, there have also been significant developments in other uses of blogs and blogging tools, and it is these in which we might glimpse the future development of the medium and understand their possible application and use in education.

Digital Identity: the online representation of an individual by themselves. Initially a "homepage" but now a complex, multi-layered presentation.

In the first instance, there is a strong move toward blogs as constructive of a digital identity. The remarkably successful *LiveJournal.com* and *MySpace.com* sites have encouraged young people to use blogs as a means of presenting themselves online, communicating with their peers, and engaging with the online world in a more holistic way than the anonymous chatroom or discussion board has ever allowed. While participants in one of these communities might communicate far more frequently through Instant Messaging (IM), email, or even SMS, it is through their blog space that they present their identity. In referring to this space through communication and in understanding that the first impression of their digital selves will doubtless be based on that space, blogs become an important indication of digital identity. Authors are not simply able to represent themselves through the content of their postings but also present much about themselves through aesthetic design, choice of media, and, perhaps most significantly, social relationships as indicative through links, comments, and other applications.

Digital identity through blogging is not, however, entirely the preserve of teenagers. As blog tools have developed, they have moved from being simply tools that allow for reverse chronological posting to providing powerful Content Management Systems (CMSs) which allow for a range of uses of the medium. In particular, this has been exploited by a rise in "professional" bloggers, who use their blogs not only to publish or discuss issues relevant to their areas, but also to create, store, and present digital portfolios. Of particular interest in an educational sense is that these bloggers are building up far more comprehensive, integrated, and engaging representations of themselves to potential employers.

In addition, blogs have, in many cases, become the focal point for particular communities. Rather than jointly owned discussion boards or community spaces, much Web-based conversation has moved onto blogs based around and facilitated through "hubs" of significant bloggers. Particularly interesting in this area is Shirky's discussion[23] of the impact of this form of communication on community behavior, highlighting as it does the decrease in "flame wars" and differing social practices (closer, perhaps, to more traditional forms of social engagement) evident in this context.

Alongside these developments are the equally significant changes not so much in writing as in reading and accessing communication through aggregation. As the statistics cited in the introduction to this collection show, blog readership in the United States and elsewhere is growing rapidly, and as of June 2005, 9 percent of U.S. Internet users surveyed by the Pew Internet and American Life Project indicated an understanding of the term RSS. Increasingly, Internet users are employing aggregation as a means of accessing information on the Web. The growth of podcasting and the incorporation of aggregation into tools such as *My Yahoo!* and personalized areas of *Google* look set to further accelerate growth in the use of this technology. Indeed, particularly significant in this case is that aggregation is used not only in connection with blogs, but rather as an entirely new mode of communication and content transfer. In an educational context, for example, previously all online communication and content management has needed to take place within specific designed environments. With the use of aggregation, this no longer has to be the case, as an individual's production can be aggregated to any number of different areas and, as is already becoming widely practiced, by individuals outside of a rigid and centralized system.

Considering these developments in the uses of blogs as digital identity and of the distributed management of information through RSS and aggregation, it is possible to conceive of an alternative to traditional models of online learning in terms of the unique qualities of these technologies.

As discussed, in a traditional OLE communication, content and participants are generally segmented into specific areas such as discussion boards, "learning modules," and synchronous chat environments. Participants are able to communicate in specified areas through bulletin boards and interact with content as separate instances. In essence, the participants are focused on the (bulletin-board) communication environments and the presented content.

However, in a blog-based OLE, while content may be accessed from a particular location it is seen to be an integral part of each blog's production through links, commentary, and more. In essence, the environment is seen to be constructed of each individual's digital identity. Further, communication between participants is centered upon each individual and facilitated through aggregation, comments on individual blogs, and the use of hyperlinks by the participants.

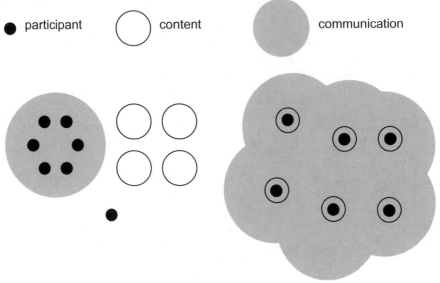

Figure 9.1:
Traditional Learning Management System

Figure 9.2:
Blog Online Learning Environment

Rather than segmented areas, courses become clusters of individuals. The capacity for participants to post to multiple categories through particular blogging tools allows them to actively belong to multiple communities, and the use of comments, email, Voice over Internet Protocol (VoIP), instant messaging, and other communication tools (frequently integrated into blog structure as alternative means of contact) allows for an extension of the interaction beyond the discussion board area. Facilitators can construct and maintain course blogs around which individuals congregate and which act as hubs, offering tools such as automated meta-course aggregation, access to key materials and resources, gradebook facilities, and so on. Rather than the environment being seen as a space in which users participate and in which communities of inquiry can develop, participants are viewed as communities of inquiry in the first instance, constructed by individuals rather than located in an abstract space.

Conclusion

While this chapter has presented models and arguments for the uses of blogs and aggregation in online education, it is important to note that these are founded on limited research and much theoretical consideration. For variants of what has been discussed to be broadly used requires not just a sympathy for the main conceptual points of discussion, but also needs much more extensive research and work in this area. It is worth considering at this point the extremely immature nature not only of the technology of blogs and aggregation, but also of all that is currently being used in online education.

However, it is not without thorough consideration of the available publications, extensive individual experience working across faculties and contexts, and a strong belief in the significance of these technologies and the medium that I draw some of these conclusions. It would seem that while there have been successful and effective uses of the discussion board, there are many significant challenges associated with its essential design, and these are exacerbated by the almost monopolistic use of the tool. Further, if (as is widely accepted) the development of a community of inquiry is seen as a valuable aim in online education, then it is the emerging provision of digital identity and alternative modes of communication through blogs and aggregation that can arguably be seen to best meet these challenges. Finally, if we are to conceive of learning as being something that happens within ourselves, and we are able to see how blogs spur this development of digital self, it is almost inconceivable that the technology will not have a very significant role to play in the development of the field.

NOTES

1. Robert Paterson, "Going Home—Our Reformation," *Robert Paterson's Weblog*, 2005, http://smartpei.typepad.com/robert_patersons_weblog/2005/02/going_home_our_.html (accessed 21 Sep. 2005).

2. Tara Brabazon, *Digital Hemlock: Internet Education and the Poisoning of Teaching* (Sydney: U of New South Wales P, 2003).

3. John Dewey, *Experience and Education* (1897; New York: Macmillan, 1938).

4. Paulo Freire, *Pedagogy of the Oppressed* (New York: Continuum, 1970).

5. *Edutools*, 2005, http://edutools.info/ (accessed 21 Sept. 2005).

6. Randy Garrison and Terry Anderson, *E-Learning in the 21st Century: A Framework for Research and Practice* (New York: Routledge, 2003).

7. Mardziah Hayati Abdullah, "The Impact of Electronic Communication on Writing" (Bloomington, Ind.: ERIC Clearinghouse on Reading English and Communication, 2003).

8. James Farmer, "Autonomy in Online (Teaching &) Learning Environments," *incorporated subversion*, 28 Apr. 2004, http://radio.weblogs.com/0120501/2004/04/28.html (accessed 21 Sep. 2005).

9. Gilly Salmon, *E-Tivities: The Key to Active Online Learning* (London: Kogan Page, 2002); Gilly Salmon, *E-Moderation: The Key to Teaching and Learning Online* (London: Kogan Page, 2000).

10. James Farmer, "Dynamic Email," paper presented at English Australia Conference, 2001.

11. Mark Guzdial, "A Catalog of CoWeb Uses," *GVU Technical Report*, 2000, GIT-GVU-00-19.

12. James Farmer, private email communication, 2005.

13. Bud Gibson, "A Learning Blogosphere," *The Community Engine*, 2004, http://thecommunityengine.com/home/archives/2005/03/a_learning_blog.html (accessed 21 Sep. 2005).

14. I. MacColl, A. Morrison, R. Muhlberger, M. Simpson, and S. Viller, "Reflections on Reflection: Blogging in Undergraduate Design Studios," *Blogtalk Downunder Conference Proceedings*, 2005, http://incsub.org/blogtalk/?page_id=69 (accessed 21 Sep. 2005).

15. James Farmer, "Centred Communication: Weblogs and Aggregation in the Organisation," *Blogtalk Downunder Conference Proceedings*, 2005, http://incsub.org/blogtalk/?page_id=54 (accessed 21 Sep. 2005).

16. Garrison and Anderson, *E-Learning in the 21st Century*.

17. David Bloom, "The Blogosphere: How a Once-Humble Medium Came to Drive Elite Media Discourse and Influence Public Policy and Elections," 2003, http://darkwing.uoregon.edu/~jbloom/APSA03.pdf (accessed 21 Sep. 2005).

18. S.C. Herring, L.A. Scheidt, S. Bonus, and W. Wright, "Bridging the Gap: A Genre Analysis of Weblogs," *Proceedings of the 37th Hawaii International Conference on System Sciences*, 2004.

19. Clay Shirky, "Power Laws, Weblogs and Inequality," *Clay Shirky's Writings about the Internet*, 2003, http://www.shirky.com/ (accessed 21 Sep. 2005).

20. Howard Gardner and Tony Hatch, "Multiple Intelligences Go to School: Educational Implications of the Theory of Multiple Intelligences," *Educational Researcher* 18.8 (1989), pp. 4–9.

21. David Squires, "Educational Software and Learning: Subversive Use and Volatile Design," *Proceedings of the 32nd Hawaii International Conference on System Sciences*, 1999, http://csdl.computer.org/comp/proceedings/hicss/1999/0001/01/00011079.PDF (accessed 21 Sep. 2005).

22. Squires, "Educational Software and Learning," p. 1.

23. Clay Shirky, "Group as User: Flaming and the Design of Social Software," *Clay Shirky's Writings about the Internet*, 2004, http://www.shirky.com/ (accessed 21 Sep. 2005).

CHAPTER TEN

Blogging to Learn, Learning to Blog

Jean Burgess

Weblogs are currently proliferating in higher education. Their emergence in Internet culture has synchronized to a large extent with trends in pedagogy toward user-centred, participatory learning in combination with the technologization of the curriculum. Weblogs are integrated into assessment as ends in themselves (in creative writing or composition courses, particularly), as formative assessment exercises, or as a means to develop critical engagement and reflection on the course material. These uses of blogging contribute to a reconceptualization of students as critical, collaborative, and creative

> **Formative Assessment:** ongoing assessment where the active demonstration of learning and cumulative feedback are emphasized. Also referred to as "assessment *for* learning," to distinguish it from **summative assessment** (or assessment *of* learning). Often includes self- and peer-assessment.

participants in the social construction of knowledge and are compatible with the social constructionist framework for learning, which—unlike the "transmission of knowledge" model—assumes that students must become active partners in the construction of knowledge *with* their peers, academic staff, and the wider social context of the disciplines in which they work.[1]

This chapter reflects on my recent experience in guiding students in the use of Weblogs as formative assessment in two university courses. Taking the social constructionist framework for learning as a starting point, the chapter reflects specifically on the problem of integrating the literacies and social communication skills that are peculiar to effective blogging into the preexisting "elaborated systems"[2] of higher education.

The first of these two courses was a cultural and media studies course focusing on popular music subcultures. The students were asked to keep a Weblog for part of the semester and to use it to discuss course content, work through and discuss their ideas for their final research projects, as well as to filter and discuss online material of relevance to the popular music culture they had decided to focus on for their research projects. The second of these courses was specifically focused on the social, political, and economic implications of contemporary "virtual" cultures, and students were asked to keep a Weblog as an online reflective journal for part of the semester.

In preparation for this task, students were given a session on how to set up and maintain a blog using *Blogger*.[3] Although there was some discussion of using a university-based solution (e.g., an open source content management system such as WordPress on a university server), because in both courses the emphasis was on engaging with Internet culture it was felt that using *Blogger* was more authentic, enabling at least the possibility of public engagement, than a "walled garden" solution. It also seemed more likely that the students' blogging practice might be sustained or repeated in other contexts, and that they might engage with the blogosphere at large if they used an open, mainstream platform.

In the popular music course, a dedicated portal, or "hub" Weblog, was established, where I not only posted a list of links to the students' blogs, news, and technical tips, but also acted as a community member, posting my own music subculture-related entries in an attempt to build interest and provide examples of good practice. In the virtual cultures course, the discussion fora within the existing standard university online teaching (OLT) system functioned similarly to this. In class, we regularly discussed the progress of the students' Weblogs—they were encouraged to reflect critically on the task and how they were integrating it into their learning and everyday lives, and we frequently explored examples of research blogging, both from within the class and elsewhere.

Blogging Literacies

Implicit in the rationale for the introduction of Weblogs as formative assessment in both of these courses was a pedagogical principle—that students should be encouraged to be active participants in the construction of knowledge—combined with a pragmatic and ethical one—that it is incumbent upon educational institutions to support students in developing literacies and competencies that are appropriate to the technological and social environments in which we all now work. Weblogs in particular appear to offer the opportunity to build the kinds of literacies that are appropriate to networked, technologized environments. These literacies can be understood as generic capabilities: increasingly essential kinds of information literacy, extending well beyond "computer literacy." We might summarize these competencies as critical, creative, and network literacies.[4]

While computer literacy is concerned with the ability to use specific hardware and software applications for instrumental purposes, *critical* technological literacy focuses on a deep, socially contextualized, and informed understanding of technology.[5] Further, the effective use of technology in educational, work, and community contexts is increasingly measured by *creative literacies*: the ability to experiment with technology in order to create and manipulate content that serves social goals rather than merely retrieving and absorbing information.[6] Most important, in the context of effective educational blogging, *network literacies* include the ability and the impulse to effectively and ethically manipulate a range of technologies to communicate and collaboratively construct and share knowledge.[7]

At first glance, these general principles of technological literacies appear to translate easily into the context of blogging in higher education. As James Farmer notes in chapter nine in this volume, blogging represents a significant move away from older, reactive "online learning" paradigms toward more productive ones. However, with this comes an expectation of greater intellectual and creative autonomy on behalf of the students, an expectation for which they are not necessarily well prepared at first, and which can take them by surprise. It is important to remember that in undertaking a university education, students for the most part engage in well-established and familiar "elaborated systems of literacy" with their attendant writing genres; blogging, on the other hand, brings with it *emergent* systems of literacy, including "new" computer literacies, and *unstable*, contested genres.

In order for students to engage effectively in the production of texts that form part of emergent genres, they need to be able to experiment and take ownership, rather than merely emulate models handed down by the instructor. Part of taking personal ownership over a Weblog, shaping its genre, and actively developing it as a space for personal learning, reflection, and interaction involves the *creative* uses of the available technologies. This requires the user to go beyond the composition of basic entries, and to extend or modify the default templates provided by the content management system or platform. Essential examples of such modifications include adding links and adding images; optional ones include changing layout or color themes. Equally, because of the ongoing and accretive practice that constitutes blogging, it is ultimately up to the students to determine for themselves their preferred format (length and style of post, frequency of posting, and depth of engagement with external materials).

However, for some students there were clear obstacles to this creative and autonomous orientation toward new technologies. It was surprising to discover the amount of time it took to teach many of the students how to insert a hyperlink in a blog entry, for example. There was often an implicit assumption by students that the technologies they use for formal learning should be stable, easy to use, and transparent, and not open, configurable, or complex. In the higher education context, this is perfectly understandable, given that their educational (as opposed to everyday) uses of technology have mostly been confined to enclosed, stable online learning platforms such as WebCT or, at most, the design of static HTML pages using, say, Macromedia Dreamweaver. While most undergraduate students these days are very comfortable using computers and the Internet for interpersonal communication (chat, email) and information retrieval (*Google*), the creative, "producerly" use of technology requires a perspectival shift that presented a significant challenge to many of the students.

The Genre Problem

As with any other "new" medium, the process through which blogging genres emerge is understood as a complex articulation of three sets of phenomena: first, the technological affordances and constraints of blogging platforms and networks; second, the ways in which the discourses around these platforms call specific user communities into being and invite specific forms of literacy, textuality, and sociality; and third, the agency of participants in shaping these communities.

Until recently, the blog has been widely understood as a coherent genre category, based primarily on a commonly agreed-upon set of formal features: reverse chronology, frequent updating, and combination of links with personal commentary.[8] However, as the discussions contained in this book show, blogging practices have diverged into a variety of directions, so that it is now necessary to understand specific genres of blogging.[9] In the context of educational blogging particularly, I would argue that in order to teach effective scholarly blogging, we need to understand the ways in which the formal and technological features of blogs combine dialogically with remediated "speech genres"[10] and existing social contexts and conventions to form hybrid sub-genres. It is not sufficient to treat a teenager's *LiveJournal*, a news pundit's blog, and a scholarly research blog as if they were merely stylistic variations on a coherent genre. In the context of educational—more specifically, "research"—blogging, existing speech genres (conversation, debate, personal storytelling) need to be articulated with ossified academic writing genres (the essay, the research report, the literature review, the critique).

Students with little or no pre-existing knowledge of the nuances of blogging genres—knowledge that can only be gained in practice—are faced with a much harder task than merely mastering the technology to the point where they can post entries. In terms of the relationship between literacy and genre, we can compare the task they face to the experience of someone who has had no formal education, but knows how to convert oral to written language, being handed a blank piece of paper and being told to write an "essay."

Brooks *et al.* have undertaken one of the more thorough investigations of genre in relation to educational blogging. For them, genres of blogging are understood as "remediations" of older writing genres (journals, essays, letters, and so on), and the purpose of their investigation was to discover which of these remediated genres motivated their students most.[11] In both of my classes, the students were directed to familiarize themselves with as many educational or research blogs as possible and to adopt whatever writing or presentation style, or mix of links, personal content, and "critical" commentary they felt was most effective, as long as they were dealing with course content.[12] Despite this, it proved extremely difficult for many students to find a writerly voice other than their most formal "essay" voice, or a personal voice other than their most casual "email" voice, for this task. It is not possible to do more than speculate on the reasons for this without further research, but it would seem intuitive to suggest that the combination of the public nature of blogging with personal ownership on the one hand, and with written assessment (which usually only reaches the "ears" of the tutor) on the other, while second nature to many full-fledged academics or commentators, puts undergraduate students in a very strange position indeed. Perhaps teaching blogging (or teaching using blogs) requires an introduction to effective *public* communication, appropriate to new media contexts, at an earlier stage.

From Textual Production to Social Practice

Moving toward considerations of blogging as social practice, Stephen Downes points out the fundamental but familiar misconception that blogging is first and foremost about writing, and suggests that we should think about blogging in terms of reading, critical thinking, and engagement with communities of practice instead:

> despite obvious appearances, blogging isn't really about writing at all; that's just the end point of the process ... . Blogging is about, first, reading. But more important, it is about reading what is of interest to you ... . And it is about engaging with the content and with the authors of what you have read—reflecting, criticizing, questioning, reacting ... the process of reading online, engaging a community, and reflecting it online is a process of bringing life into learning.[13]

In a distinct shift away from the tradition of the scholarly essay with which the students were familiar, the approach taken to the Weblog task in these two courses emphasized the social, networked nature of knowledge construction and sharing, rather than the production of particular kinds of written texts. That is, at least ideally, this "scholarly" form of blogging was understood as social, and not only textual, practice. Further, in my case, the genre expectations I had of the students cannot be separated from the course content. Because both the music subcultures course and the virtual cultures course had a heavy emphasis on social and cultural contexts and networks, so, too, I emphasized the importance of adopting a network orientation, rather than simply a "writing" orientation, to the blogging task. This meant an emphasis and an expectation, reflected in the marking criteria, on direct commenting, linking to and discussing classmates' work and found resources or other online discussions, and integrating personal experiences as a part of other online or offline communities into the content of the Weblog.

Lessons Learned

In both courses some students took to blogging like ducks to water, while others were bemused, reluctant, or downright hostile to the idea. Some students immersed themselves in online culture for the duration of the task and made the blog a seamless part of their identity as students, while at the other end of the scale, some students simply summarized readings or lectures, ignoring their blogs and those of their classmates until the next long, link-free entry. The journey toward finding a balance between a critical engagement with the literature, meaningful use of the network, and engagement with readers and other writers *on* the network is not an easy one, and if I am honest, it is a balance I still struggle with in my own blogging practice.[14]

But, interestingly, one of the more surprising lessons learned through these (relatively) early experiments in blogging for formative assessment was that, if we give students a voice, they will most certainly use it. In the popular music course, one of the most vehement objectors to the blogging task took the time to put together a post of several hundred words critiquing the pedagogical philosophy behind compulsory student blogging.

Generally, I found that those students who were already curious, outgoing, and interactive in small-group tutorials transferred those competencies—which, it should emphasized, are *social* competencies, not only "academic" ones—onto the blogging task. It was these students who were more likely to seek out the blogs of their colleagues and leave links, and who engaged most actively with the offline or online cultures they were studying. This meant that the blogs kept by these more socially active students were the most effective, both in the specific contexts of their media and communications disciplines *and* in the context of the blogosphere.

Possibly as a result of a repeated emphasis in the guidelines and in class on the need to do more than write posts, I found that most students were reasonably active in both direct commenting and distributed networking (linking and commenting on others' work in their own blogs). In the popular music course, I also posted entries on the "hub" blog to draw attention to interesting posts on student blogs, and I spent time each week reading and commenting on individual student blogs where I felt it was appropriate. Although the commenting tended to cluster around pre-existing social groups in the class, those who made the effort to visit and read each other's blogs and leave comments were rewarded by inclusion in these emerging networks, and in semi-formal evaluations at the conclusion of the semester, several students noted that the formation of learning communities with other students outside individual small-group tutorials was an unexpected benefit of participating in the blogging task. It is also true to some extent that international students, particularly those for whom English was not their first language, were far more active participants in the distributed, online mode of discussion made possible by the use of Weblogs than they were in traditional in-class discussions.

On reflection, it seems clear that students in both of these courses who integrated their blog into their everyday lives (e.g., by posting short, reflective entries almost daily, or by integrating the theory with their personal interests) also engaged more effectively with, and went beyond, the course content. From this it is possible to draw the conclusion that the effective integration of the personal, social, and formal aspects of scholarly blogging increases the learning value of blogging, especially when used as formative assessment.

Conclusions

It has to be acknowledged that the cases discussed above represent a somewhat artificial and contrived exercise, particularly in that, unlike blogging for English composition courses, the requirement to focus on unit content (music subcultures in one case, issues around "virtual cultures" in the other) makes the challenge of representing an authentic, dynamic, and engaged self through a course Weblog all the more difficult: put simply, personal research Weblogs work best when they focus on the intellectual passions of their creators, and some students will only ever be as superficially invested in subject-specific Weblogs as they are in those subjects in the first place. However, Weblogs can also amplify the effects of learner engagement: the more that students are encouraged to connect theory to their own experiences, the more effective the course Weblog is as a space for learning. It is therefore important from the outset to encourage students to choose research topics that match their personal experiences and interests. Finally, the evidence from the two courses discussed here lends itself to arguments in favor of portfolio-based assessment. That is, if students were required or encouraged to keep a Weblog for the duration of their degree, rather than starting and then dropping individual course Weblogs one by one, it seems more likely that they would treat their Weblogs as authentically social spaces that are more meaningfully representative of them as individual learners, community members, and cultural citizens, thereby increasing student engagement with, and ownership of, the learning process.

If this kind of critical and creative engagement with learning is to be constituted effectively through blogging, students and teachers need to better understand how to build the critical, creative, and network literacies that are required for, and are in turn enhanced by, effective educational blogging. These matters are far from straightforward, and a thoroughgoing engagement with them represents a major future challenge to research into the uses of technology in education. Much of the existing literature, while engaging in significant depth with the ways in which Weblogs can be integrated into the curriculum and the affordances they offer,[15] is still in the exploratory and explanatory stages. I would argue that there are major advantages in progressing our understanding of these issues to be gained by viewing blogging as a complex field of *cultural* and *social* practice—at least as much as, if not more than, we understand it as textual production. The payoff will be learners who are both more engaged with the world around them, and better equipped to be active, literate, and critical participants in an increasingly networked and technologically complex world.

NOTES

1. Thomas Angelo, *Doing Academic Development as Though We Valued Learning Most: Transformative Guidelines from Research and Practice*, Paper presented at the HERDSA Annual International Conference 1999, Melbourne.

2. In this context, the term "elaborated system" has resonances with both linguistics and the formalist tradition of literary studies, and is used to describe the ways in which "ideal" communicative forms become stable within equally stable genre systems, in relation to particular social and institutional contexts.

3. *Blogger*, recently acquired by *Google*, is arguably the most widely used Web-based blogging platform. It can be found at http:// blogger.com/.

4. Some of the theoretical framework in this chapter was developed in association with Jude Smith, Ross Daniels, and Axel Bruns as part of the research for a large teaching and learning grant at Queensland University of Technology. The grant project eventually developed a framework addressing a set of critical, creative, and collaborative ICT literacies. The idea of "network literacy" is inspired by Jill Walker, for whom it means "linking to what other people have written and inviting comments from others … understanding a kind of writing that is a social, collaborative process rather than an act of an individual in solitary." See Jill Walker, "Weblogs: Learning in Public," *On the Horizon* 13.2 (2005), pp. 112–18.

5. National Research Council, *Being Fluent with Information Technology* (Washington, DC: National Research Council: Commission on Physical Sciences, Mathematics and Applications; Committee on Information Technology Literacy; Computer Science and Telecommunications Board, 1999).

6. Mark Warschauer, *Technology and Social Inclusion: Rethinking the Digital Divide* (Cambridge, MA: MIT P, 2003).

7. Franklin Becker, Kristen L. Quinn, and Carolyn M. Tennessen, *The Ecology of Collaborative Work* (New York: International Workplace Studies Program, Cornell University, 1995); Richard Davis and Dennis Schlais, "Learning and Technology: Distributed Collaborative Learning Using Real-World Cases," *Journal of Educational Technology Systems* 29 (2001), pp. 143–56; Robert Godwin-Jones, "Emerging Technologies: Blogs and Wikis: Environments for On-Line Collaboration," *Language Learning & Technology* 7.2 (2003), pp. 12–16.

8. Probably the most widely cited definitions of blogging genres are to be found in Rebecca Blood, *The Weblog Handbook: Practical Advice on Creating and Maintaining Your Blog* (Cambridge, MA: Perseus, 2002).

9. See also Carolyn R. Miller and Dawn Shepherd, "Blogging as Social Action: A Genre Analysis of the Weblog," in Laura J. Gurak, Smiljana Antonijevic, Laurie Johnson, Clancy Ratliff, and Jessica Reyman, eds., *Into the Blogosphere: Rhetoric, Community, and Culture of Weblogs*, June 2004, http://blog.lib.umn.edu/blogosphere/blogging_as_social_action _a_genre_analysis_of_the_weblog.html (accessed 27 Oct. 2005).

10. Mikhail M. Bakhtin, *Speech Genres and Other Late Essays*, 1st ed., trans. V.W. McGee (Austin: U of Texas P, 1986).

11. In general, studies of educational blogs and their uses in composition classes frame blogging as writing, not as social practice, although there has been more consideration of audiences and publics for blogging recently. See, for example, Charles Lowe and Terra Williams, "Moving to the Public: Weblogs in the Writing Classroom," in Gurak *et al.*, *Into the Blogosphere*, June 2004, http://blog.lib.umn.edu/blogosphere/moving_to_the_ public.html (accessed 27 Oct. 2005).

12. Kevin Brooks, Cindy Nichols, and Sybil Priebe, "Remediation, Genre, and Motivation: Key Concepts for Teaching with Weblogs," in Gurak *et al.*, *Into the Blogosphere*, June 2004, http://blog.lib.umn.edu/blogosphere/remediation_genre.html (accessed 27 Oct. 2005).

13. Stephen Downes, "Educational Blogging," *EDUCAUSE Review* 39.5 (2004), pp. 14–26.

14. My own Weblog, *Creativity/Machine* (http://hypertext.rmit.edu.au/~burgess) is a personal research Weblog. It was begun as a way of collating material I was collecting for my Ph.D. proposal, but has evolved into a space for discussion, networking, and dissemination of my research.

15. For a major contribution to such work, see Jeremy B. Williams and Joanne Jacobs, "Exploring the Use of Blogs as Learning Spaces in the Higher Education Sector," *Australasian Journal of Educational Technology* 20 (2004), pp. 232–47.

SECTION TWO
BLOGS IN SOCIETY

CHAPTER ELEVEN

Scholarly Blogging:
Moving toward the Visible College

Alexander Halavais

S cholars who blog are engaging in more than personal publishing; they are shaping a new "third place" for academic discourse, a space for developing the social networks that help drive the more visible institutions of research. The number of blogging scholars and the novelty of the medium mean that what blogging is and how it relates to being a scholar in the networked age remains unresolved, but the inchoate informal networks of blogging scholars that exist today already hint at the potential of the practice.

New technologies inevitably draw on earlier models to make sense of how they should be used, and to offset the potential social disequilibrium brought about by the technology.[1] We are in the midst of a quiet, uneven revolution in academic discourse, and blogging and other forms of social computing make up an important part of that revolution. We may filter our view of blogging through a set of archetypal scholarly communication settings: the notebook, the coffee house, and the editorial page. For now, scholarly blogs are a bit of each of these, while they are in the process of becoming something that will be equally familiar, but wholly new.

Bias of Blogging

As noted in earlier chapters, so varied are the behaviors of bloggers that it is a bit surprising that the same term is used to cover them all. Nonetheless, there are four themes that seem to form a core set of practices and beliefs among many bloggers. First, blogs rely on networked audiences that may share little in common except for being regular or irregular readers of a particular site. Mass media act to collect audiences and aggregate opinion and attention; blogs encourage individualized views of the informational world.

> **A-List Blogs:**
> those blogs that attract the largest readership and are frequently hyperlinked. The exact makeup of the A-list is disputed, but services like *Technorati.com* help track the most often linked blogs.

A second hallmark of blogging is that it encourages conversation. Often commentators have focused on so-called "A-list" blogs, which many not value exchange as highly. Other bloggers might be classified as "mumblers"—without obvious comments or readers. Even in these cases, though, it seems that bloggers are seeking a way of conversing with the world.

Third, blogging is a low-intensity activity. Producing microcontent requires little commitment of time, and free blogging platforms provide an inexpensive outlet for this microcontent.

Finally, blogs represent a relatively transparent and unedited view of thinking-in-progress.

While there are examples of Websites using blogging software that do not exhibit all four characteristics, they are accepted broadly enough to constitute a bias of the medium, a tendency of practice. It is not difficult to find antecedents to these overall themes in both the history of hacking and of scholarship—two cultures that share significant common ground.[2] A decade ago Harrison and Stephen explained why computer networking was of such interest to academics. It played to long-held ideals among scholars that had yet to be realized: "unending and inclusive scholarly conversation; collaborative inquiry limited only by mutual interests; unrestrained access to scholarly resources; independent, decentralized learning; and a timely and universally accessible system for representing, distributing, and archiving knowledge."[3] Blogs, while not addressing all of these ideals, have already shown themselves to be effective in ways that other, centrally organized efforts at scholarly networking have not.

The Notebook

Perhaps no tool is more closely associated with scholarly pursuits than the notebook or journal. It represents a first attempt to externalize knowledge and ideas. C. Wright Mills, in *The Sociological Imagination*, extols the virtue of a particular kind of notebook:

> in such a file as I am going to describe, there is joined personal experience and professional activities, studies under way and studies planned. In this file, you, as an intellectual craftsman, will try to get together what you are doing intellectually and what you are experiencing as a person. Here you will not be afraid to use your experience and relate it directly to various work in progress. By serving as a check on repetitious work, your file also enables you to conserve your energy. It also encourages you to capture "fringe-thoughts": various ideas which may be by-products of everyday life, snatches of conversation overheard on the street, or, for that matter, dreams.

Once noted, these may lead to more systematic thinking, as well as lend intellectual relevance to more directed experience.[4]

The notebooks of many scholars, from Faraday to da Vinci to Gramsci to Darwin, have opened up new realms to later researchers.[5] While the use of such a notebook differs from field to field, in sciences it might be argued that the lab notebook represents a clear expression of everything the scientist does.[6] The notebook represents an externalization of the investigator's memory and cognition.[7] Even in the case of diaries, which are often considered to be very private, a public audience—if only potential—is always posited in some form.[8] In fields where collaboration is the norm, notebooks are actively shared, and this aids in the social process of thinking about a problem.

From the earliest conceptions of a networked notebook, the idea that they may be accessible by a wider community of like-minded individuals has been central. Vannevar Bush's hypothetical "memex ... an enlarged intimate supplement to [the individual's] memory," provides one example:

> the owner of the memex, let us say, is interested in the origin and properties of the bow and arrow ... . First he runs through an encyclopedia, finds an interesting but sketchy article, leaves it projected. Next, in a history, he finds another pertinent item, and ties the two together. Thus he goes, building a trail of many items. Occasionally he inserts a comment of his own, either linking it into the main trail or joining it by a side trail to a particular item.[9]

These "trails" of associative links, Bush suggests, will be available to the owner of the memex for future use, and to other researchers. More recently, electronic notebooks are shared within "collaboratories."[10]

For some scholars, a blog replaces the notebook as a way of externalizing thought. Science fiction author Cory Doctorow refers to his blog as an "outboard brain" and notes that the process of creating a note in such a way that an audience will understand the importance of a link or idea "fixes the subject in [his] head the same way that taking notes at a lecture does, putting them in reliable and easily-accessible mental registers."[11] Social cognition generally thrives on common understandings and experiences among those communicating, but Doctorow suggests that by making his ideas as plain as possible to a wider audience, it allows him and others to better think about the problem.[12]

Blogs seem to be particularly good at establishing and exchanging what Merton calls "specified ignorance."[13] While there exist other venues for presenting work in progress, ideas published on blogs often draw immediate responses. This unaccustomed immediacy makes some scholars feel like "kids in a candy store."[14]

Specified Ignorance:
Robert Merton's term for "a new awareness of what is not yet known or understood and a rationale for its being worth the knowing."

Opening up and documenting intellectual work also allows for apprenticeship. Because a record, or "trail" as Bush put it, is publicly available on the Web, anyone may learn from it. Documented conversations in other contexts allow individuals to come to terms more quickly with the norms of a particular discourse community,[15] but because blogs are often records of what individuals are doing in the "real world," they provide the same ability to someone wanting to acquire a set of knowledge practices.[16]

The blog as a research notebook is an effort to move thought into the social realm, by presenting facts, ideas, and requests for assistance—and ultimately build knowledge[17]. Scholarship is always communal, and scholarly communication has continually moved toward quicker and more interactive forms, but never have so many had access to so many, so easily. The transparency of blog content draws affinitive hyperlinks and readers who comment, and that ultimately leads to collaboration. As Nardi, Schiano, and Gumbrecht note, "readers create blogs as much as writers."[18]

The Coffee House

The coffee house of eighteenth-century London provided a setting for free exchange of ideas among nearly any interested parties. Many have suggested that blogging is the modern recapitulation of pamphleteering, but like blogs, the coffee house thrived on mixing and exchanging the opinions and ideas of those from a variety of backgrounds. Like blogs, however, those with particular interests or political leanings were likely to flock to the same coffeehouse.[19]

Bloggers establish loose communities by linking their blogs through comments, reactions, and hyperlinks. Such discourse and linkage communities may not be common in the larger blogosphere, and many bloggers may not participate in them.[20] One of the difficulties in identifying scholarly blogs is that many of them do not link together. Nonetheless, there are gathering places for academic bloggers where topics of particular concern to scholars remain a common part of the conversation.

Public Sphere: most often associated with Jürgen Habermas and his *Structural Transformation of the Public Sphere*, the public sphere constitutes a sphere where private individuals can come into open dialog to form public opinions.

One way of detecting such discourse communities is to trace hyperlinks among them. While conversations utilize other channels beyond hyperlinks, the link network represents a rough trace of such conversations. Figure 11.1, for example, represents a network of reciprocal links between scholarly bloggers.[21] Other approaches to detecting such conversations are rapidly emerging, and visualizing these conversations on a larger scale provides an opportunity to understand how these structures differ from place-based communities like the historical coffee house.[22]

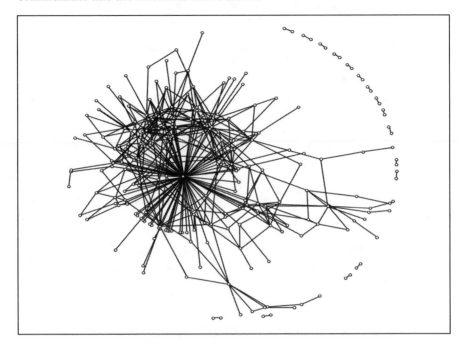

Figure 11.1: Network of 253 scholarly blogs with reciprocal hyperlinks, surrounding the *Crooked Timber* group blog

In blogging circles, a single blogger is unlikely to see the broad discussion as a whole, while the physical constraints of the coffee house made the extent and shape of discussion clearer. Despite this significant difference, the transparency and ease with which bloggers can enter the conversation provides a replacement for the idealized form of the coffee house.

The Opinions Page

The newspaper inherited the legacy of the private journal and the coffee house, creating an organ of mass public communication. The role of literary journals and scholarly journals more generally has been central to the development of intellectual life. But academics have often entered the broader public arena, presenting ideas within and outside their own area of expertise. There has been a long history of exchange between intellectuals and journalists, and blogs have extended this.

Umberto Eco, who has written both scholarly work and popular fiction, has also written regular newspaper columns.

I don't believe there is any gap between what I write in my "academic" books and what I write in the papers. I cannot say precisely whether, for the papers, I try to translate into language accessible to all and apply to the events under consideration the ideas I later develop in my academic books, or whether it is the opposite that happens. Probably many of the theories expounded in my academic books grew gradually, on the basis of the observations I wrote down as I followed current events.[23]

A disproportionate number of scholarly bloggers come from fields that are most directly related to public policy: law, economics, and political science, among others; though as more scholars blog, the range of fields increases. For scholars who feel it is their responsibility to educate the public and engage in public issues, blogs lower the barriers to communicating directly to the public.

Scholarly blogging in practice often includes some portion of all three of these models. Mortensen and Walker note that "when we started our blogs, we saw them mainly as tools for focusing, for exchanging information, and being part of a conversation which potentially extends beyond the academic community ... our blogs became tools with which to think about our research, its values, connections and links to other aspects of the world."[24] As scholars engage the practices of blogging, we are likely to change those practices and be changed by them.

Impediments to Scholarly Blogging

Blogs seem to fit the existing needs and ideals of scholars, and wide adoption might spur significant improvements in scholarly communication. The role of the intellectual in society is predicated on the ability to interact with two groups: a wider public that is interested in and provides reputational support to the scholar, and a group of peers who may provide a structure of exchange and review, both of which blogging appears to offer.[25] While there are a relatively larger number of bloggers among the professoriate and university students, given the advantages of blogging, why do we not see more scholars who blog?

Pierre Bourdieu extends the idea of capital, suggesting that while scholars may not be wealthy in economic terms, they express their structural position through cultural preferences.[26] Cultural capital is afforded to them (in part) through the institutions in which they engage. Popular opinion often sees blogging as faddish and trivial. Simply the fact that anyone can easily start a blog makes is suspect from the perspective of the cultural elite.[27] Moreover, the structures of affirming academic positions, especially in the United States, require scholars to behave in a particular way: to "publish or perish." Despite the very low barriers to entry, creating and maintaining a blog leads to opportunity costs, both culturally and economically.

The blogosphere provides its own intrinsic reputational rewards, but these may not extend to the wider academic (and other) contexts in which scholars work and are valued. The cumulative value of both citations to blogs and senior scholars taking up the practice of blogging may yield slow change in research institutions. However, given that the embrace of electronic publishing has been "just around the corner" for decades, and has only made significant inroads in physics and some of the other natural sciences, it seems unlikely that many institutions will place their imprimatur on blogging as a practice in the near future.

One reason for this is that blogging so often mixes the personal with the professional. This may already be common in the academic world, where work and home are often interpenetrated in space and time. However, such interpenetration does not necessarily mean that the two are combined. The transparency of blogging, especially when authors are identified by name, leads to an unusual collapsing of the public and private sphere, a regression to rural life and concentric social circles.

The very elements of blogging that make it most valuable—a networked audience, open conversation, low barriers to entry, and transparency—are also most threatening to established strictures of academic behavior. While each may be valued by individual scholars, the university as an institution in many cases relies on treating the public as a mass, providing authority to limited channels of communication, constructing barriers to scholarly discourse, and maintaining bureaucratic partitions between academe and other parts of the life of a scholar. A recent essay in the *Chronicle of Higher Education* warned academics seeking jobs to avoid blogs precisely because of these properties, and the possibility that a hiring committee may be threatened by a candidate's blog.[28] The claim was answered by a blogger and academic, Henry Farell, who concludes that despite the unruly and novel nature of blogging, it provides "a kind of space for the exuberant debate of ideas, for connecting scholarship to the outside world, which we haven't had for a long while."[29]

Scholarly Blogs and Visible Colleges

Scholars will continue to blog, but what practices will help to differentiate "scholars who blog" from "scholarly blogging"? Blogging will remain a part of scholarly life that exists outside of academic institutions, even as it is practiced from within them. Scholarly blogging will thrive in two ways: first as informal discourse

The Invisible College: this term refers to the collective creation of a school of thought by a distributed group of scholars, often using both formal and informal channels to communicate their ideas.

community, and second in articulation with existing values and structures of academe.

While invisible colleges are usually analyzed through published literature, Crane notes that the "social circles" that underlie such structures look very similar to those we have discussed above, with a fluctuating membership that is often aware of some, but not all, of the other members.[30] Geographic separation means that face-to-face contact in the same location is not possible. Likewise email is not a common form of communication among such loosely affiliated networks, tending to be used largely by those who are already friendly.[31] Electronic media have supported invisible colleges in the past, but they have always played a secondary role to face-to-face meetings. Blogs provide an electronic "third place," to use Oldenburg's term, a public space in which people are able to interact with one another, free from the necessity of introductions.[32] The improvements Crane identifies in communication for invisible colleges are linked to making such communication more visible and accessible.[33] Blogs provide an electronic version of the coffee house and the academic conference, allowing for open and observable discussion and debate among near-strangers.

In this "third place" role, blogs largely support rather than replace traditional institutions and channels of communication. The informal communication that occurs on blogs helps to raise interesting questions, forge collaborative work, and involve scholars in a larger public sphere. Existing institutions and processes may find challenges from blogs; the transparency of blogging complicates double-blind review, for example. But generally, blogging provides scholarly communication that otherwise would be missing, a virtual place for continual discussion with little cost or commitment from participants. While blogging may not garner the sort of academic recognition that traditional channels do, as with other forms of creating professional relationships, the informal networks supported by blogging provide the foundation required to excel within the institutions of academe. For this reason, over the long time it takes academic institutions to change, there will be a growing recognition of the place of blogging in the scholar's life.

NOTES

1. Carolyn Marvin, *When Old Technologies Were New* (New York: Oxford UP, 1988).

2. Pekka Himanen, *The Hacker Ethic* (New York: Random House, 2001).

3. Teresa M. Harrison and Timothy Stephen, "Computer Networking, Communication, and Scholarship," *Computer Networking and Scholarly Communication in the Twenty-First-Century University* (Albany: State U of New York P, 1996), p. 32.

4. C. Wright Mills, *The Sociological Imagination* (New York: Oxford UP, 2000), p. 196.

5. See, for one example, Ryan D. Tweney, "Faraday's Notebooks: The Active Organization of Creative Science," *Physics Education* 26 (1991).

6. Marcello Sarini, Enrico Blanzieri, Paolo Giogini, and Claudio Moser, "From Actions to Suggestions: Supporting the Work of Biologists through Laboratory Notebooks," *Proceedings of the COOP2004 Conference*, French Riviera, May 11–14, 2004.

7. Andy Clark and David Chalmers. "The Extended Mind," *Analysis* 58(1) (1998).

8. José van Dijck, "Composing the Self: Of Diaries and Lifelogs," *Fibreculture Journal* 3 (2004), http://journal.fibreculture.org/issue3/issue3_vandijck.html (accessed 27 Oct. 2005).

9. Vannevar Bush, "As We May Think," *The Atlantic* 176.1 (1945).

10. Daniel C. Edelson, Roy D. Pea, and Louis M. Gomez, "The Collaboratory Notebook," *Communications of the ACM*, 39(1) (1996).

11. Cory Doctorow, "My Blog, My Outboard Brain," *O'Reilly DevCenter Articles* (2002), http://www.oreillynet.com/pub/a/javascript/2002/01/01/cory.html (accessed 27 Oct. 2005).

12. Robert M. Krauss and Susan R. Fussell, "Constructing Shared Communicative Environments," in *Perspectives on Socially Shared Cognition*, Lauren B. Resnick, John M. Levine, and Stephanie D. Teasley, eds. (Washington, DC: American Psychological Association, 1991).

13. Robert K. Merton, "Three Fragments from a Sociologist's Notebooks: Establishing the Phenomenon, Specified Ignorance, and Strategic Research Materials," *Annual Review of Sociology* 13 (1987).

14. Quoted in David Glenn, "Scholars Who Blog," *Chronicle of Higher Education* 49(39) (6 June 2003), p. A14, http://chronicle.com/free/v49/i39/39a01401.htm (accessed 27 Oct. 2005).

15. See Thomas Erickson, "Social Interaction on the Net: Virtual Community or Participatory Genre?" *SIGGROUP Bulletin*, 18(2) (1997).

16. Lilia Efimova, Sebastian Fiedler, Carla Verwijs, and Andy Boyd, "Legitimized Theft: Distributed Apprenticeship in Weblog Networks," *Proceedings of I-KNOW04*, Graz, Austria, 2004.

17. Sébastien Paquet, "Personal Knowledge Publishing and Its Uses in Research," *Seb's Open Research*, 3 Oct. 2002, http://radio.weblogs.com/0110772/stories/2002/10/03/personalKnowledgePublishingAndItsUsesInResearch.html (accessed 27 Oct. 2005).

18. Bonnie A. Nardi, Diane J. Schiano, and Michelle Gumbrecht, "Blogging as a Social Activity, or, Would You Let 900 Million People Read Your Diary?" *Proceedings of the 2004 ACM Conference on Computer Supported Collaborative Work*, 2004.

19. Aytoun Ellis, *The Penny Universities* (London: Secker & Warburg, 1956).

20. Susan C. Herring, Inna Kouper, John C. Paolillo, Lois Ann Scheidt, Michael Tyworth, Peter Welsch, Elijah Wright, and Ning Yu, "Conversations in the Blogosphere: An Analysis 'From the Bottom Up,'" *Proceedings of the 38th Annual Hawaii International Conference on Systems Sciences*, 2005.

21. Alexander C. Halavais, "Schoolblog Network," A *Thaumaturgical Compendium*, 6 June 2004, http://alex.halavais.net/news/index.php?p=728 (accessed 27 Oct. 2005).

22. For examples of such attempts to map conversations, see Eytan Adar, Li Zhang, Lada A. Adamic, and Rajan M. Lukose, "Implicit Structure and the Dynamics of Blogspace," *Workshop on the Weblogging Ecosystem, 13th International World Wide Web Conference*, May 18, 2004; Brian M. Dennis and Azzari C. Jarret, "NusEye: Visualizing Network Structure to Support Navigation of Aggregated Content," *Proceedings of the 38th Hawaii International Conference on System Sciences*, 2005; and Lilia Efimova and Aldo de Moor, "Beyond Personal Webpublishing: An Exploratory Study of Conversational Blogging Practices," *Proceedings of the 38th Hawaii International Conference on System Sciences*, 2005.

23. Umberto Eco, *Travels in Hyperreality* (New York: Harvest Books, 1986), p. x.

24. Torill Mortensen and Jill Walker, "Blogging Thoughts: Personal Publication as an Online Research Tool," *Researching ICTs in Context*, Andrew Morrison, ed. (Oslo: InterMedia Report, 2002).

25. For a discussion of the context of the intellectual in society, see Lewis A. Coser, *Men of Ideas* (New York: Free Press, 1965).

26. Pierre Bourdieu, "Forms of Capital," *Handbook of Theory and Research for the Sociology of Education*, J. G. Richardson, ed. (Westport, CT: Greenwood Press, 1986).

27. Torill Mortensen, "Personal Publication and Public Attention," *Into the Blogosphere: Rhetoric, Community, and Culture of Weblogs*, Laura J. Gurak, Smiljana Antonijevic, Laurie Johnson, Clancy Ratliff, and Jessica Reyman, eds., June 2004, http://blog.lib.umn.edu/blogosphere/personal_publication.html (accessed 27 Oct. 2005).

28. Ivan Tribble, "Bloggers Need Not Apply," *Chronicle of Higher Education* 51(44) (8 July 2005), p. A3, http://chronicle.com/jobs/2005/07/2005070801c.htm (accessed 27 Oct. 2005).

29. Henry Farrell, "The Blogosphere as a Carnival of Ideas," *Chronicle of Higher Education* 52.7 (7 Oct. 2005), http://chronicle.com/free/v52/i07/07b01401.htm (accessed 27 Oct. 2005).

30. Diana Crane, *Invisible Colleges* (Chicago: University of Chicago Press, 1972).

31. Emmanuel Koku, Nancy Nazer, and Barry Wellman, "Netting Scholars: Online and Offline," *American Behavioral Scientist* 44(10) (2001).

32. Ray Oldenburg, *The Great Good Place: Cafes, Coffee Shops, Bookstores, Bars, Hair Salons, and Other Hangouts at the Heart of a Community* (New York: Marlowe & Co., 1999).

33. Oldenburg, p. 121.

CHAPTER TWELVE

Blogging from Inside the Ivory Tower

Jill Walker

I 've found this article almost impossible to write. You see, I used to love blogging. I've blogged for over five years. I've written papers about blogs. I've taught classes with blogs. I've blogged about completing a Ph.D., teaching my first classes, getting a real job and becoming the head of my small department. I must have given a few dozen talks on blogs, to students and professors and librarians and artists and school children and engineers and teachers. You'd think it would be easy to write a book chapter about research blogs. And yet I've found this article almost impossible to write.

I find it harder to blog, too, to be honest. When Torill Mortensen and I wrote what appears to have been the first research paper on blogs,[1] I was an outsider trying to enter the ivory tower. We were graduate students in a new and only half-accepted field in a country on the outskirts of the world: of course we loved blogging. Blogging allowed us to circumvent the power structures of academia and geography. We found our voices. We heard ourselves, we heard each other, and we were heard by others. It was exhilarating.

Today I blog and my students read the post. I post a photo to *Flickr* and the next time I log in I notice that ten new people have made me their contact since last week, at least half of them students, I think, but I'm not sure. I meet a professor in another department for the first time and he says, lifting an eyebrow, "I've read your blog," and I know too little about him to be able to interpret that eyebrow. I preferred blogging when most people didn't know about it.

When I started writing this chapter, I thought I could construct a serious article discussing different kinds of research blogs and exploring the virtues of the genre. I read research discussing how scholars use communication technology. I wanted to see research blogs as a step in the long history of academic discourse. And yet every time I began to write, I felt that anxiety I thought frequent blogging had dispelled long ago: the anxiety that if I said what I really wanted to say, I wouldn't know how to defend it. My blog mirrored my disease, and my posts became briefer, less open, and less frequent.

I've tried to write this article as a serious, academic discussion, and it hasn't worked. This essay insists on being about something more personal: why is blogging no longer easy? Why was blogging so immensely liberating when I was a Ph.D. student, and yet so complicated now? What happens when research blogs—and their authors—become part of the academic system rather than being outsiders trying to get in?

The Importance of Hierarchy

Back when I still thought I could write this essay in a conventional scholarly manner, keeping my ambivalence and anxiety about blogging neatly outside of the text, I read research on scholarly mailing lists, which rose in importance during the eighties and nineties. Despite the enduring success of some mailing lists, most have devolved into distribution channels for conference announcements or for irregular and off-topic random postings. Blogs are still on the ascendancy and may never go through such a decline, but veterans of mailing lists still invariably compare blogs to this slightly older genre.

An interesting, if unpleasant, reason for the decline of useful, constructive discussions on mailing lists has been suggested by Timothy Stephen and Teresa M. Harrison in their 1994 study of *Comserve*, an electronic community. Stephen and Harrison found that the relative anonymity and openness of mailing lists dissolves the hierarchical systems that are necessary for the academic system to work:

> in a well-known case study, Zuboff (1988) documented the tension created within a corporation when a computer-based electronic communication system was installed. The openness, inclusiveness, and anonymity of computer-mediated communication was antithetical to the organization's hierarchical structure; it facilitated the rise of democratic dialogue among workers, thereby placing stress upon traditional hierarchical roles.[2]

This democratic dialogue is exactly what Torill Mortensen and I praised in our "Blogging Thoughts" paper. We described an example where a professor's article about bloggers was met by vociferous and critical responses from the bloggers themselves. The bloggers' reaction was strong enough to require a response from the professor—however, he didn't have a blog of his own, so could not respond in the same space as the bloggers who were discussing his work. Instead, he posted a response in one of his students' blogs, in an unusual inversion of the usually hierarchical access to the media.[3]

I still love these inversions of conventional power relationships that blogs and other forms of online publication make possible. And yet Zuboff, as cited by Stephen and Harrison, notes in particular the *tension* that this inversion creates. Is the unease I feel in writing this article and in blogging at all these days related to that tension? I would rather not think so, but objectively it's clear that while my conventional academic status is not nearly as high as that of the professor Torill Mortensen and I wrote about, I'm now very clearly positioned within the conventional academic hierarchical system. Perhaps blogging disrupts hierarchies that now serve me well, and that makes me ambivalent about them?

Stephen and Harrison continue:

> As Holzner (1968) has described, many work communities define themselves by regulating the contexts in which work occurs and by imposing strict controls on the flow of communication. Academic work communities are no different; indeed, many of Holzner's examples are drawn directly from the sciences.
>
> Thus, while it is easy to be enamored of the networks' potential for enhancing democratic exchange, it is worth considering that the culture of the academy is essentially an hierarchical meritocracy. Voices become privileged through individual accomplishment; in fact, the degree to which communication is restricted is often one of the few public signs of a scholar's disciplinary stature. Those at the top of a disciplinary hierarchy are more difficult to access and often restrict their public communication to prestige channels. Thus, it is doubtful that achievement-oriented academics will be natively inclined to carry their dialogue to a venue in which the relative anonymity of authors and audiences reduces the ability to gauge the impact of one's contribution. One might anticipate that there would be a tendency for academics to restrict their communication on the public network channels to information of lesser consequence.[4]

In many ways, blogs support the academic hierarchy better than mailing lists. While a scholar posting to a mailing list can add a signature below the text of his or her email to establish credentials, those who are not already aware of the participants in the field and their place in the hierarchy of experts are unlikely to fully grasp this. In contrast to mailing lists, traditional academic conferences have far higher entry thresholds as well as clear methods for signaling who has the most clout. For one thing, the most important people within the community will be on stage, giving the keynote or other high-profile lectures, and they'll spend most of their time off the podium speaking to each other rather than with the plebs or grad students.

Most scholarly mailing lists are open to anyone who is interested, although subscribers generally do have to sign up to read the messages. Blogs are even more easily accessible. Readers do not have to sign up and can often participate in the discussion without any special membership.

On the other hand, an individual blog functions in much the same way as the lecture podium. One person, or a group of people, is up on the podium and is thus very clearly defined as the main speaker, the person you should be listening to. Questions and comments from the audience are often encouraged, but they are clearly positioned as subordinate to the main speaker's words.

So is my ambivalence about blogging—and about writing this essay about research blogging—caused by my having bought into the academic system? On the contrary, I think my blog, and the blogs of other academics, are clearly positioned in the academic hierarchy. My blog, as many other academic blogs, gives my name as Dr. Jill Walker, it states the department I work at and my university, the "about" section gives my title, my research interests, and the courses I'm teaching. I suppose I might unconsciously be trying to restrict communication to emphasize a higher status, as Stephen and Harrison suggest, but that would seem counterproductive, given the positive responses blogging has generally brought me in academia, such as invitations to write papers, give talks, and teach classes.

It's more complicated than that.

What Is a Research Blog?

There are several different kinds of research blogs, some of them closer to the traditional forms of academic publication than others. These work in different ways. Let me describe a few of the kinds.

Public Intellectuals

Many academic bloggers use their blogs as a platform for political debate based on theories of political science, feminism, discourse and media analysis, and so on. *Bitch PhD* is an example of a political blog (though it is not exclusively political) that discusses current events and personal experience in a theoretical and generally feminist manner.[5] *Crooked Timber* is another example, being a group Weblog of a number of political scientists.[6]

Research Logs

The "pure" research log is a record of research conducted and ideas that might be pursued. There are several different traditions for this. Humanists have traditionally kept notebooks to collect quotations and ideas, laboratory scientists have kept records of their experiments, and engineers and architects draw sketches of designs. The dissertation log might be seen as a special case of research log, especially when written specifically for a supervisor and not for a general audience.

Pseudonymous Blogs about Academic Life

The kind of title given to this proliferating branch of the academic blog is characterized by a tongue-in-cheek refusal to revere the ivory tower experience:

Dr. Crazy, *Confessions of a Community College Dean*, and *Barely Tenured* are but a small sampling of titles chosen.[7] Being pseudonymous, these researchers generally don't reveal their exact field of specialization, but instead tend to discuss more general aspects of academic life and the paraphernalia of research. While not exactly research blogs, these blogs certainly allow the discussion of issues more likely to be discussed in the coffee breaks than the presentation sessions of a conference: how to find the time to do research, how to behave at a conference, the process of earning a Ph.D. or tenure, and so on.

These different genres of academic blogs form such separate clusters that authors in one genre show little awareness of the existence of the other genres, rarely cross-linking and sometimes even complaining at the lack of other genres.

When I began blogging, I intended to write a pure research log. What I ended up writing was a hybrid genre, my favorite kind of research blog: the blog that both discusses the content of research, the ideas themselves, and that also discusses the process and experience of researching. I think to explain that better, I need to tell you the story of my experience as a blogger.

Becoming a Blogger

I began blogging on the day I discovered *Blogger.com*: October 9, 2000. I was reading *Caterina.net*, and I can't remember how I was there or whether I was conscious that this was a "blog," but I do remember the moment I saw the blue and orange button in one of the sidebars: "Powered by Blogger." I think I clicked it more out of vague curiosity than with any clear idea that this was something that would change my life. Yes, I know, that sounds rather extreme, doesn't it? Change my life. But in many ways it did, you see.

Blogger.com didn't have the same design then as it does now. Writing this, I took a look at *Archive.org* to see what, precisely, I saw on that October morning. The blue and orange was the same as now, but most of the page was actually devoted to the company blog. The most recent posts, when I saw the page for the first time, celebrated the creation of the ten thousandth blog on *Blogspot*, *Blogger*'s still-existent free hosting service, and noted the publication of a *New Yorker* interview with Meg Hourihan, one of the founders of Pyra, the company that developed *Blogger*. There were also links to the most recently updated blogs and, most important, there was the slogan, the thing that ensured I was a blogger ten minutes later: "Push-button publishing for the people." In big letters just below, it said: "Create your own blog! Blogger offers you instant communication power by letting you post your thoughts to the web whenever the urge strikes."

I was still three years away from successfully defending my Ph.D., a young would-be scholar in a nascent field sitting in a coastal town in Norway a long way from most of the world. Of course I clicked the Start Now! button. Wouldn't you?

As soon as I'd started my own blog, I began looking for other blogs. I only found one other Norwegian blog (written in English) and no other research blogs. Of course, back then, I don't suppose I really knew what a research blog was, or could be. But soon enough I found myself writing about my work (easy to do since my research was on new media narrative and art) and asking researchers I'd met at conferences to start blogs of their own. Many did.

Blogs, after all, are inherently social. Whether you have five readers or five hundred doesn't really matter; it's the knowledge that this will be read that is important. For a very new scholar, used to a world where three people will read a paper, grade it, and then put it away forever, that is an exhilarating and frightening idea.

Over the next years, my blogging was an important tool in my research. I deliberately used my blog to develop my writing voice, just as one might keep a journal as a tool to improve one's writing and thinking. The most important way my blogging helped my research was social, though. Through my blog I developed and connected to a network of other people, mostly Ph.D. students but also people in industry and more experienced academics who were interested in the same topics as I. There weren't many other research bloggers back then. In 2002 I compiled a list of all the research blogs I could find, and reached about twenty, though I found a few new ones every week. By 2003 I couldn't keep up with the emails from new research bloggers asking to be included, and I gave up trying to keep the list up to date. Today Crooked Timber's list of academic blogs includes several hundred blogs, and that list is far from comprehensive. Lilia Efimova and Stephanie Hendrick have shown how Weblog clusters change over time, often spurred by events like conferences where bloggers meet colleagues, merging online and offline worlds.[8]

About the same time, having blogged for a year or two, my colleague Torill Mortensen and I wrote our paper on research blogs, "Blogging Thoughts."[9] We wrote with enthusiasm about the promise of blogging and the ways in which blogging had been a valuable tool in our research. Torill and I both still blog, years later, and though we probably would have written that article a little differently today, we still pretty much agree with what we wrote. And yet things have changed.

Today I'm a tenured scholar, I chair a small department, I advise grad students and plan curricula and apply for funding and carry out all the other myriad tasks a professional scholar performs. It's impossible to say objectively whether my blogging helped me reach that goal. I know that my blogging helped me gain a foothold among researchers in my field, that the regular writing and discussions with readers and other bloggers helped me become a confident writer, and that I had more opportunities to give talks and write in other genres than most of my non-blogging peers. So, quite probably, blogging helped me succeed in earning a Ph.D. and getting my first academic job.

Does blogging still help me? I'm not so sure. Now my students read my blog, my colleagues read my blog, and quite possibly (though I tend to assume not) the department secretary and the dean read my blog. No doubt my daughter will read some of it when she's older. Five years ago, I could safely assume none of my colleagues would read my blog unless they were especially Internet-savvy and actually interested in the topics I write about.

Blogging does not allow for the changes in roles we're used to in different relationships. I behave differently when I'm camping with my family or having Sunday brunch with friends than when I'm in a meeting with the dean or discussing the budget with my colleagues. I tell my girlfriends different and more complete versions of my life than I tell my students. Most perplexing are the strangers, people I've never met, or whom I might have taught in a single class. Although 99 percent of my life is unblogged, these people think that they know me. I used to laugh when they asked me how I could stand people knowing all about me. I stopped laughing last year. I was talking intimately with a girlfriend in a foreign city at a bus stop, assuming we had the privacy of a public place far from home. A few weeks later I noticed a link to my blog from a blog written by a woman in the city I'd visited. She wrote that she'd heard two women talking by a bus stop and had realized that it was *jill/txt* and a friend and that she knew just what they were talking about. That is a side-effect of blogging that had never occurred to me, and that makes me want fewer readers, not more.

A discomfort with putting forward one face to strangers, friends, and colleagues has little to do with research blogging specifically. These issues are general problems in blogging, and different bloggers solve them in different ways. The other discomfort I've recently felt with blogging is more directly due to being perfectly accepted within the traditional academic system.

Fellow bloggers told me that one of the reasons they'd enjoyed reading my blog as I was finishing my Ph.D. is that I included a lot of posts about the *process* of research and specifically about completing such a large and daunting task. Comments in my blog and in emails told me that it had been useful to them to read daily about someone going through the same thing, just a few months or years ahead of them.

It's different, though, being on the inside of the system. Most of the process of academic life on the inside of the system is in fact intensely interesting to academics. We lap up blogs that go through the details of the actual life of research and teaching: *Cheeky Prof, Barely Tenured, Just Tenured, New Kid on the Hallway, Bitch PhD, Wanna Be PhD, Learning Curves*; all these and more blog their to-do lists (grade papers, write report, prep teaching, finish manuscript, feedback to X, prep meeting), their frustrations and joys with students or colleagues or conferences, and the many ways in which their personal lives intersect with the academic profession. *Confessions of a Community College Dean* even succeeds in writing about university administration in an entertaining and engaging manner.[10] These blogs are precisely about the process of research and academic life rather than about results or precise topics.

Notice, though, that each and every one of those blog titles cloaks identity. They're all pseudonymous. Some academic blogs are written using a real name, in which case they'll probably be focused on traditional "content" and research, although there is also a strong genre of political academic blogs, generally written (unsurprisingly) by political scientists. Others, just as important, are pseudonymous, where you get all the honest process work and the bits that are too bodily (sex! mess! clothes! hunger!) or emotional (performance anxiety, depression, love, doubt) to fit into a traditional academic image.

It is striking that the popular genre of pseudonymous academic blogging completely rejects academic hierarchies by refusing to identify the blogger. Or perhaps it would be more accurate to say that pseudonymous bloggers embrace nothing but position in the hierarchy: almost all state the position of the blogger as adjunct, grad student, dean, tenure-track professor, recently tenured assistant professor, etc. Perhaps this allows them to explore the tension created by the openness and democratic lack of hierarchy of these networks, noted by Stephen and Harrison above, rather than to be controlled by it.

Research Practice

There is, then, a split between pseudonymous blogs and blogs where the author writes under her or his real name. Pseudonymous blogs tend to write about the process of research and teaching but leave out the content and documentation of research, that which has traditionally been the province of academic publications. Real name blogs tend to stick either to political issues not closely connected to the blogger's research, or else they document research by linking to the author's publications but include little discussion of the process of blogging. Graduate students' blogs are often the most successful at combining the process and the "content" of research.

Blogging is writing *practice*, Kathleen Fitzpatrick recently wrote. She was comparing regular blogging to the daily practice musicians are accustomed to, pointing out that the more she blogged, the more she was able to blog:

> when I discipline myself to post something every day, or as close to it as I can, I find myself watching the world around me slightly differently, and treating my thoughts slightly differently, as though any occurrence or any idea might be capable of

blossoming and bearing fruit. When I'm not posting, nothing seems worth writing about, just a bunch of dried-up seeds that'll eventually blow away or be eaten by the birds.[11]

Research blogging can likewise be research practice. As Lilia Efimova has pointed out, research bloggers rarely explicitly document their research in their blogs.[12] Although I agreed with her on this at first,[13] after a little thought I changed my mind:

> Yesterday I agreed with Lilia that most researchers' blogs don't document research. Today while reading a post on David Weinberger's blog I realized that that's completely beside the point: research happens in blogs, and in the conversations between blogs. Blogs aren't about documentation, they're about doing, thinking and discussing. And they're about catching fleeting thoughts and making them explicit: if I hadn't blogged my response to Lilia yesterday I probably wouldn't have thought about David's post today as research and wanted to rethink yesterday's ideas as I'm doing now.[14]

Traditionally, the process of research has been transient: conference discussions, conversations with colleagues over coffee, and reviewers' reports on unpublished manuscripts. Only the final publications remained, usually with traces of uncertainty, false starts, considerations, and process neatly edited out. For a Ph.D. student, the academic system has built-in mechanisms to help with the process of research and to make it visible. There are specific graduate seminars, you meet regularly with an advisor, and you're expected to accept correctives from senior academics. You're an advanced learner, but still very clearly a learner. As an academic in a permanent or tenured job at a university, your position changes. Now there are few role models for making the process, the *practice* of research visible. Small wonder, then, that while graduate students often blog openly and exploratively about their own research, most professors seem to blog either pseudonymously, rarely, or relatively impersonally, in almost all cases sticking to blog topics that have little to do with their own research practice.

I do still believe in blogging. I think that we have yet to find the real place of blogging. Whether or not blogging still exists in its current form in ten years' time, the public sharing of research practice is likely to become more and more visible. We'll have to deal with it somehow. Hopefully we'll avoid raised eyebrows and people in foreign cities knowing too many of our intimate secrets.

Will We Write Academic Papers in 2035?

Scholars work in the genres of their time. Socrates did his academic work and dissemination in dialogues with his students. Print publication is only a necessity of scholarship in today's book-bound world. Today many universities measure our publication rates in carefully weighted systems directly tied to our funding. Is this really going to be the way of the future?

One way of looking at Weblogs and emerging forms of scholarly discussion and work is that they are the popularization of research, or a new form of dissemination. If they allow ideas to be worked through, it is in the same way as informal conversations in the breaks at scholarly conferences do, or perhaps at best they can replace or augment the debates that ideally (though usually not really) take place in the question sessions after traditional scholarly papers are presented.

What if the future of scholarship is not in papers and books, but in new forms of dialogue-based exploration of ideals? Could professors in 2035 use some descendant of Weblogs as their site of developing ideas *and* as the main form of dissemination?

Weblogs in their current form can't fully replace traditional publication. They're superficial, quotidian; they're not rigorous enough, one might argue; they are too completely in the moment and encourage fast writing and thought rather than deep consideration and reflection. And yet it is obvious that bloggers tend to revisit the same issues again and again. Many bloggers are adept at linking back to related entries written months or even years earlier, both by themselves and by others. The link itself has become something of an ethics of blogging: link to your sources. If you're not sure of a fact or of the source of your ideas, search the Web until you find out more about that and link to it. These foundations are, perhaps, the seeds of a genre that may grow to be as strong as the traditional academic essay.

The other great advantage of traditional publication over blogging, for the moment, is duration and accessibility. True, it is often easier to google something that is blogged rather than to find an academic paper, possibly not even online, that discusses the same issue. But the traditional academic paper is guaranteed longevity by a well-designed library system. A blog discussion trail just a year or two old is hard to follow through its broken links and shut-down servers. Being able to put things on bookshelves can help—or at least it can help those who have access to the bookshelf.

What would an academic community after print look like? Ted Nelson, the man who coined the word "hypertext" in the sixties, once said he'd "never imagined the techies would try to simulate *paper*,"[15] and yet that is exactly what most online journals do: they publish traditionally written and formatted papers as PDFs. Some journals, such as *Kairos*,[16] are experimenting with essay formats that integrate links, images, video, and exploratory techniques that really use the medium. And yet even these essays remain clearly defined, whole objects.

Weblogs have no whole; they are not objects. They are processes, actions, sites of exchange, more like Socrates's original dialogues must have been than Plato's written version of them.

Will Weblogs develop into something that is both scholarship in action and a complete form of dissemination and storage of ideas for the future? Or will we always need a Plato to write down these momentary dialogues in a way that can be stored for posterity?

We'll have to wait and see. In the meantime, let's keep exploring new genres, pushing the limits, and thinking about the kinds of tools we would need to make the future for scholarship that we truly want.

NOTES

1. Torill Mortensen and Jill Walker, "Blogging Thoughts: Personal Publication as an Online Research Tool," in Andrew Morrison, ed., *Researching ICTs in Context* (Oslo: InterMedia, University of Oslo, 2002), pp. 249–79.

2. Timothy Stephen and Teresa M. Harrison, "Comserve: Moving the Communication Discourse Online," *Journal of the American Society for Information Science* 45.10 (1994), pp. 768–69.

3. Mortensen and Walker, "Blogging Thoughts," p. 263.

4. Stephen and Harrison, "Comserve," pp. 768–69.

5. See http://bitchphd.blogspot.com/.

6. See http://www.crookedtimber.org/.

7. See http://crazyphd.blogspot.com/, http://suburbdad.blogspot.com/ and http://maplestreet.blogs.com/.

8. Lilia Efimova and Stephanie Hendrick, "In Search for a Virtual Settlement: An Exploration of Weblog Community Boundaries," 2005, https://doc.telin.nl/dscgi/ds.py/Get/File-46041 (accessed 3 Nov. 2005).

9. Mortensen and Walker, "Blogging Thoughts."

10. See http://suburbdad.blogspot.com/.

11. Kathleen Fitzpatrick, "Practice, Practice, Practice," *Planned Obsolescence*, 11 Sep. 2005, http://www.plannedobsolescence.net/index.php?/weblog/practice_practice_practice/ (accessed 3 Nov. 2005).

12. Lilia Efimova, "Why Weblogs Are Rarely Used to Document Research (2)," *Mathemagenic*, 16 April 2003, http://blog.mathemagenic.com/2003/04/16.html#a555 (accessed 3 Nov. 2005).

13. Jill Walker, "Documenting Research," *jill/txt*, 5 June 2003, http://jilltxt.net/?p=182 (accessed 3 Nov. 2005).

14. Jill Walker, "Not Documenting, Doing," *jill/txt*, 6 June 2003, http://jilltxt.net/?p=184 (accessed 3 Nov. 2005).

15. Ted Nelson, "Fixing the Computer World," *Incubation* 3, Nottingham (2004).

16. See http://english.ttu.edu/kairos/.

The Political Uses of Blogs

Mark Bahnisch

When media commentary refers to blogs, often this is shorthand for political blogs.[1] There is no doubt that, increasingly, a perception exists that blogs are heavily involved in the political sphere,[2] as participants in agenda setting, in launching critiques of public policy, in interfacing with election campaigns, in influencing political debate and events, and in sparking activism. For instance, *The Economist* referred in an article to the heat of speculation in the blogosphere about the (then) forthcoming indictment of Dick Cheney's Chief of Staff, "Scooter" Libby.[3] The paper regularly refers to the influence of the blogosphere in discerning the feelings of the Republican base, such as the reaction of activists to the nomination of Harriet Miers to the Supreme Court.[4]

There has been little academic research to date on political blogs and the politics of blogging, probably because the phenomenon is relatively novel. Rather than presenting a survey of the literature, or a report on empirical research, therefore, this chapter takes a more theoretical and speculative approach to questions regarding the place and influence of political blogs in an increasingly fragmented and dispersed political and informational culture. Nevertheless, the chapter will address these questions in an iterative way through reference to the blogosphere.[5]

In assessing the contribution made by political blogs to the political process and to civic engagement, the chapter offers some thoughts on two hypotheses that are implicit in much debate and commentary about blogs.[6] The first relates to the degree to which blogs can contribute to the quality of political conversations. It is now commonplace in political sociology to observe that the ideal-typical public sphere of classical liberal political theory—where truth emerges from rational debate among active citizens and its reflection through the media and political representation[7]—is an increasingly hollowed-out space.[8] Concentration of media ownership, corporate agendas for news setting, and the limiting of open debate both in political fora and in public conversation mirror secular declines in citizens' interest and participation in politics. For liberal democratic theory, such shifts pose a major challenge, along axes of concentration of power and the rationality—as opposed to partiality—of political debate. It has been argued that the blogosphere can offer a corrective to the decline in the respect for truth in public debate, and thus contribute to a revival of the public sphere.[9] This argument needs to be critically examined.

Related to the first hypothesis is the framing of blogs in terms of their purported ability to replace, supplant, or supplement the role of professional journalism in enforcing public accountability and in influencing political commentary.[10] This argument is normally developed in two ways. The first is an assertion that the interactivity of blogs makes them more attractive to a public increasingly distrustful of formerly authoritative sources of information. The second is the contention that blog commentary—through the passion that is said to characterize it—is more attractive to readers than the increasingly ritualized and often hackneyed commentary of old media. These arguments also need critical examination. From a sociological perspective, it is appropriate to be somewhat skeptical of sweeping generalizations based on a short-term lens and little empirical evidence as to the novelty of phenomena such as online political interaction and commentary. However, this is not to say that there may not be some truth in both hypotheses, even if the contribution that blogs can make to politics and to public debate and activism may prove to be more modest than claimed by both their advocates and detractors.

The chapter proceeds first by discussing the history and current contribution of political blogs to public debate and politics,[11] and next through assessing the state of the political blogosphere against the two hypotheses outlined above. Finally, some conclusions will be drawn, and suggestions for future research advanced.

The Political Blogosphere

Political blogs rose to prominence in the United States—with blogs like Andrew Sullivan's[12] on the right and the *Daily Kos*[13] on the left—gaining a wide audience in 2001 and 2002. The first major controversy in which blogs were seen to have taken a role revolved around the remarks that Senate Majority Leader Trent Lott (R-Miss.) made on the death of long-serving former senator Strom Thurmond (R-S.C.). Thurmond's political career was remarkable. As Governor of South Carolina, he had opposed Truman's re-election in 1948; as the candidate of the States' Rights Democratic Party, he had been elected to the Senate in 1954 on a write-in vote; and he had switched to the Republicans as the impact of the Civil Rights revolution grew in 1963. Thurmond retired from the Senate at the age of 100, as its longest-serving senator. American politics can be seen as profoundly shaped by the Civil Rights movement, both in terms of the hegemony gained by the Republican Party over the formerly solidly Democratic South, and also in terms of the increasing attention given to cultural and identity issues at the expense of economic and class cleavages. Lott's downfall—aided and abetted by the investigative work of journalist Josh Marshall at his blog *Talking Points Memo*[14]—related to the perception that he had implicitly endorsed the racism that characterized Thurmond's early career through his comments regarding Thurmond's presidential race. It was significant that the first (and very high-profile) "scalp" taken by the blogosphere was associated with two broader and very significant trends in American politics: the increasing partisanship that has characterized the permanent campaign since the 1990s, and the central node of the culture war—race. Marshall's role was not substantially different from that of a traditional investigative journalist (indeed, that is his professional background), and there is more resonance with the post-Watergate phenomenon of close scrutiny of political figures. But what was significant about this event was the fact that what would otherwise have been a little-noticed speech was placed firmly on the agenda of the mainstream media, and kept there through the immediacy and impact of blog writing.

Subsequent years saw the mainstreaming of the blogosphere—with assistance from the Iraq War and the presidential primaries and general election of 2004. Blogs were increasingly read and discussed by political consultants and staffers, journalists, and politicians, and have to some degree become part of the political landscape. *Campaigns Online*[15] picked up at an early stage on the influence that blogs were having on gaining support and disseminating messages in the primaries—particularly in the (unsuccessful) campaign of Democratic former governor of Vermont Howard Dean, but to a more limited degree in other primary campaigns and in the general election race of Senator John Kerry (D-Mass.). Bloggers were credentialled at party conventions, conservative bloggers claimed a scalp in Rathergate, and both news reports and op-ed articles began to take the pulse of the blogosphere, as well as appropriating particular themes and issues. *Wikipedia* reports that by the end of 2003, the three most prominent American blogs—the conservative *Instapundit*[16] and liberal blogs *Atrios Eschaton*[17] and *Daily Kos*—each had more than 75,000 unique visitors per day. Some idea of the influence on the election that such blogs were perceived as having can be discerned from this contemporary post, as well as some insights into the hierarchical nature of the blogosphere, a theme that will recur:

> Who plays this role on the blogs, of listing the zeitgeist, and having daily "push" into the rest of the blogspace?
> Two bloggers do: Atrios on the left and Instapundit on the right. Nearly every political blogger reads one of them, or both, every day. Both basically list links to other blogs and newspapers, with commentary spliced in. Instapundit never breaks stories, Atrios only occassionally; the value they impart is aggregating the most compelling stories of

the day into a blog feed, and providing the spin, either liberal or conservative, onto those stories. ...

Atrios and Instapundit each have upwards of 30,000 readers a day. Though small in audience when compared to major media outlets, theirs is an influential readership. The DNC, Howard Dean (and every candidate with a blogroll) links to, reads, and quotes Atrios. Instapundit frequently spotlights stories, like the emergence of Iraqi bloggers, that are then picked up by newspapers and online publications such as Townhall.com. Among the readership of both blogs are journalists, press critics, cable pundits, and opinion-makers. The liberal think-tank Center for American Progress draws from Atrios; the anonymous Atrios is even jokingly accused of being Sidney Blumenthal, the former Clinton aid[e].

In the blogspace, both Atrios and Instapundit help direct the general conversation. Aside from the influential reach of their audience, both Instapundit and Atrios are read by thousands of small bloggers, who often follow the linked stories and add their own commentary. A link from either Atrios or Instapundit will result in thousands of visitors to a site. In order to garner a link, bloggers will tailor their writing to what Atrios or Instapundit wants to have featured. The influence of these mega-bloggers thus radiates outward; what they talk about is what the blogs are talking about. This important positioning is self-sustaining. Like the gossip blog Gawker, each essentially operates by getting tips from their readership; both receive hundreds of emails a day from bloggers hoping to receive a link, or just from commenters or readers who email tips.[18]

The contemporary perception of the influence of the blogosphere is aptly captured by Mallory Jensen, writing in the *Columbia Journalism Review*:

the growing power of Weblogs, or "blogs," has hardly gone unnoticed. Bloggers have been credited with helping to topple Trent Lott and Howell Raines, with inflaming debate over the Iraq war, and with boosting presidential hopeful Howard Dean. Suddenly, it seems, everyone from Barbra Streisand (whose site is a lefty clearinghouse) to guy-next-door Bruce Cole (a San Francisco foodie whose blog is called Sauté Wednesday), has been swept into the blogosphere.[19]

Another very significant trend of note in the American blogosphere is the degree to which blogs can have a financial impact on elections, and the related phenomenon of the professionalization of blogs and bloggers. *Daily Kos* founder Markos Moulitsas[20] can serve as an exemplar. *Daily Kos* channeled half a million dollars in reader donations to Democratic candidates in 2004, purportedly forcing changes to electoral strategy by Republicans. Moulitsas—along with other "A-list" American bloggers—makes a living from blogging, and has been tapped by campaigns as a consultant, by the British newspaper *The Guardian* as a guest blogger, and has a book contract.[21] Blogs have also become a resource for expert and informed commentary through academic blogging with a political tinge, exemplified by Juan Cole[22] and group blog *Crooked Timber*.[23] Such commentary is disseminated both across the blogosphere and into the media and the political and policy communities, utilizing the sort of interface that is not dissimilar to that of the traditional role of the academic as public intellectual, but much more immediate.

As with most matters Internet, attention in the media and the limited academic literature has largely focused on American blogs. Canadian blogs have a wide influence:

> Weblogs of a political nature have a high visibility in Canada, perhaps more so than other countries. The current Prime Minister kept a high-profile blog during his campaign for leadership of the Liberal Party. Opposition Finance Critic Monte Solberg keeps a widely read blog, as does former advisor to Jean Chrétien, Warren Kinsella. The contents of political blogs have been used to both defend and attack politicians in Parliament.[24]

Canada also exemplifies another trend in the blogosphere—increasing aggregation[25] through group blogs or portals (such as *Progressive Bloggers*[26] or *The Blogging Tories*[27]) along partisan lines.

The Australian blogosphere provides a good contrast with the American environment, as it exemplifies some of the same trends, but also displays some differences related to political and media culture. Blogs are also grouped along ideological lines—ranging from *Larvatus Prodeo*,[28] *Back Pages*,[29] *The Road to Surfdom*,[30] and *John Quiggin*[31] on the Green and Social Democratic Left, through centrist blogs such as *Troppo Armadillo*[32] and the largely libertarian *Catallaxy*,[33] through to the right-wing blog that dominates the political right of the blogosphere—the eponymous *Tim Blair*.[34] Australian blogs have not been as partisan in party terms as American ones[35]—a distinction that probably reflects the more open nature of American political parties and may also be related to the large number of academics and economists represented among Australia's prominent political bloggers, as opposed to the activists and journalists who dominate the largest American blogs. With some exceptions, journalists are much more dismissive of the contribution and promise of the blogosphere,[36] which no doubt reflects the quality and insularity of the Australian media. Blogs have played more of a role in commenting and providing a sense of community in elections—most notably on the left, *Back Pages*[37]—rather than direct participation and agenda-setting. Again, a less freewheeling civil society in the political culture and the significantly shorter time span of Australian elections would account for some of this contrast. Australian bloggers perceive some influence on the shaping of debates in the public sphere,[38] and as Christopher Sheil argues, the poor and tired quality of comment and opinion will no doubt lead to a more direct challenge in the future.[39] In part because of the small size of the Australian media market, there is as yet little professionalization of the blogosphere, though that can also be predicted to change, with the same trends toward aggregation and hierarchization visibly emerging[40] as in Canada and the United States.

Discussion

Although the review of the history and current influence of blogs in this chapter does not pretend to be exhaustive or definitive, it is sufficient to establish that blogs are influencing politics and the political process. Some *caveats* need to be added—the blogosphere tends to reflect the mainstream media in focusing more on the partisan and opinion-forming aspects of public debate rather than on input into the policy process, but that may be something that will develop over time. Particularly with the expert input that academic and professional bloggers can bring, this is an eminent possibility—recognizing that such bloggers may already be policy players. Similarly, bloggers are able—both through the lack of constraint on topics compared to the mainstream media and also through their ability to weave threads that unify themes that are repeated and iterated[41] over time—to play a real part in framing political discussion, arguably in a more sustained way than the traditional op-ed. As attention increasingly focuses on the construction of narratives that reveal new dimensions of the political,[42] the influence of the blogosphere, both in dissemination of political narratives and in an interplay with other political actors, will no doubt grow. This sort of contribution—being able to publish reflective and controversial analysis that would rarely find its way onto the op-ed pages, and being able to follow up quickly on developments—is a real service that the blogosphere performs for democracy. This is arguably more important than "breaking stories." Much of the debate over the contribution of blogs has been misframed, in particular in terms of the debate about blogs and the mainstream media—bloggers as the new journalists, blogs as the new op-ed, blogs as the new way for politicians to interface with voters. All this actually acts to obscure the most important feature of blogging—its ability to engage directly with readers and generate continuing and iterative political conversations.

As argued above, one *leitmotif* in political theory and commentary is the decline of the public sphere and the disconnection between the increasingly professionalized sphere of party politics and the mass of citizens. Much of this harks back to the liberal (in the philosophical sense) model of truth as emerging from debate and discussion. In fact, on blogs as elsewhere, much political debate—whether between Left and Right or along other cleavages—is marked by missed encounters, a failure to engage, sniping, snarkiness, and spin. This chapter has already noted repeatedly the increase in partisanship and aggregation along ideological axes that are clear trends in the blogosphere. Although there is some continued conversation across ideological lines,[43] it is difficult to discern within the blogosphere anything greatly approaching the pure public sphere of reason lauded as an ideal type by Habermas. The increasing imbrication of bloggers (particularly in America, and with the media and publishing industries and mainstream political networks) also raises the Weberian question[44] of the degree to which the originally amateur phenomenon of blogging will come to mimic the behavior of those who "live off politics" rather than for it. However, these *caveats* need also to be understood in an appropriately tentative and measured way, as pointing to possibilities and avoiding value judgments.

In all of this, blogging reflects not just a broader decline in civility, but something about the very nature of political discourse: it is not about getting to the truth but about swaying others through means fair and foul.[45] Similarly, in the masculine tone of many political blogs, and the dominance of male commenters and bloggers, blogs are very much embedded in and part of their environing society. There is also often a confessional dimension to blogging, as Michel Foucault argues of late modern culture more broadly. Blogging is not without its "look at me" egotism and its own internal hierarchies and exercises of power that structure the conversations that take place and the relative influence and deference with which ideas and arguments can be articulated. Most characteristically, as with the Internet generally, already-existing social hierarchies are often reinforced (though perhaps shifting in other ways according to aspects of the medium and the subjectivities of blogging), particularly those related to class and gender. It would be of great interest to do a discourse analysis of the adoption and reproduction of classed and gendered speaking positions in the blogosphere. It is also worth noting the degree of inter-cultural misunderstanding that can arise,[46] and pre-eminently the erection of new hierarchies based on hits, links, memes, and A-lists, which are peculiar to the blogosphere but reflective of contextual cleavages and power relations.

Where blogging may represent a distinctive contribution to political debate, however, is in the ease with which such hierarchies can be challenged. That is not to say that they will be. Ironically, but predictably, from a sociologist's point of view, it would seem that the very features and developments that have brought blogging to greater prominence, and that give the promise of increasing political influence, may be those that reduce its distinctiveness as a political phenomenon and in particular its ability to challenge entrenched forms of unequal political interaction and power relations.

Blogging is a new platform for political persuasion, and for making a sustained argument over time in a way that the mainstream media rarely do, and bloggers can be directly challenged as to facts and interpretation. If prominent Australian blogger Tim Dunlop is right that bloggers are the new public intellectuals,[47] then the difference is that they are directly accountable—in public—to their own peers and readers. The other major difference is that in an age of declining public interest in politics, comments threads (very often the best bits of blogs, for the sharpness and wit of the encounters and the sense that different commenters get of each other's personas) and the interactivity and virtual community they bring with them are a small protest against the disconnect between citizens and political process. As much as media and political account-ability and sharp analysis are important, interactivity (and the political interest and involvement it fosters) is perhaps the key contribution blogging brings to politics and to society. As blogger Ken Parish put it, bloggers and commenters become "monitorial citizens."[48] Just as this chapter has argued that blogging reflects broader social patterns, so, too, it may be that this political interactivity is a sign of the times. Conversely, the question must remain open as to the de-gree of future influence blogs can offer that is distinctive and a value addition to other forms of political and social action. Strong reservations remain about the envisioning of the blogosphere as an ideal(ized) public sphere.

Conclusion and Directions for Future Research

This chapter has examined the history and significance of political blogs through a theoretical context informed by contemporary social theory and po-litical sociology. The chapter concludes that blogs represent a very interesting, and potentially quite distinct, contribution to politics, but expresses some de-gree of skepticism as to whether the factors leading to their rise may also lead to their colonization by social relations of inequality, which are political in a different sense. However, further research is urgently needed, particularly in mapping the reach and influence of blogs, and also in a more rigorous and empirically informed analysis of conversations and power relations internal to the blogosphere and their relationship to their environing contexts. Neverthe-less, claims that blogs are a revival of a pure liberal public sphere or a quantum leap beyond the increasingly commodified mass media are very much over-drawn. However, there is no doubt that the ways blogs have shaped politics to date, and have the potential to shape politics in the future, represent possibili-ties that are potentially productive of liberatory outcomes.

NOTES

1. *Wikipedia*, "Political Blog," 2005, http://en.wikipedia.org/wiki/Political_blog (accessed 16 Nov. 2005).

2. Tim Dunlop, "We Hates Bloggers, We Hates Them," *The Road to Surfdom*, 27 Oct. 2005, http://www.roadtosurfdom.com/archives/2005/10/trevor_cook_get.html (accessed 16 Nov. 2005); John Quiggin, "Walkley on Blogs," *JohnQuiggin.com*, 22 Oct. 2005, http://johnquiggin.com/index.php/archives/2005/10/22/walkley-on-blogs/ (accessed 16 Nov. 2005).

3. "The Plame Scandal: Santa Is on His Way," *The Economist* (22 Oct. 2005), p. 34.

4. Sue MacDonald, "The Big Picture on a Red-Letter Day," *Blogpulse Newswire*, 28 Oct. 2005, http://blog.blogpulse.com/archives/000441.html (accessed 16 Nov. 2005).

5. The author maintains a political blog at http://larvatusprodeo.redrag.net/ and has been an observer and participant in the Australian political blogosphere in recent years. For this reason, impressions gained, except where the context indicates otherwise, will primarily be sourced from Australian blog and political culture, with reference also made more specifically to U.S. examples, as the American blogosphere is recognized as the most influential and extensive instance of the broader phenomenon.

6. For ease of reading, references to "blogs" in this chapter can be taken to refer only to political blogs, except where explicitly indicated otherwise.

7. Jürgen Habermas, *The Structural Transformation of the Public Sphere: An Inquiry into a Category of Bourgeois Society* (Cambridge: Polity Press, 1989).

8. Kate Nash, *Contemporary Political Sociology: Globalization, Politics, and Power* (Oxford: Blackwell, 2000).

9. Tim Dunlop, "If You Build It They Will Come: Blogging and the New Citizenship," *Evatt Foundation*, 15 Nov. 2005, http://evatt.labor.net.au/publications/papers/91.html (accessed 16 Nov. 2005).

10. David Higgins, "Power to the People," *Walkley Magazine*, 2005, http://magazine.walkleys.com/index.php?option=content&task=view&id=20 (accessed 16 Nov. 2005).

11. For reasons of space, and because of the limited literature identified above, this discussion does not purport to be definitive but rather presents key snapshots of the development and influence of the blogosphere. The discussion is in part indebted to the *Wikipedia* entry at http://en.wikipedia.org/wiki/Blog#Blogging.27s_rise_to_influence.

12. See http://www.andrewsullivan.com/.

13. See http://www.dailykos.com/.

14. See http://www.talkingpointsmemo.com/.

15. Alexis Rice, "The Use of Blogs in the 2004 Presidential Election," *CampaignsOnline.org*, Oct. 2003, http://www.campaignsonline.org/reports/blog.pdf (accessed 16 Nov. 2005).

16. See http://www.instapundit.com/.

17. See http://atrios.blogspot.com/.

18. Matt Stoller, "The New AP: Atrios versus Instapundit, Part One," *Bopnews*, Jan. 2004, http://www.bopnews.com/archives/000120.html (accessed 16 Nov. 2005).

19. Mallory Jensen, "Emerging Alternatives: A Brief History of Weblogs," *Columbia Journalism Review*, 2003, http://www.cjr.org/issues/2003/5/blog-jensen.asp (accessed 16 Nov. 2005).

20. *Wikipedia*, "Markos Moulitsas Zúniga," 2005, http://en.wikipedia.org/wiki/Markos_ Moulitsas_Z%C3%BAniga (accessed 16 Nov. 2005).

21. *Wikipedia*, "Daily Kos," 2005, http://en.wikipedia.org/wiki/Daily_Kos (accessed 16 Nov. 2005).

22. See http://www.juancole.com/.

23. See http://www.crookedtimber.org/.

24. *Wikipedia*, "Progressive Bloggers," 2005, http://en.wikipedia.org/wiki/Progressive_ Bloggers (accessed 16 Nov. 2005).

25. See also Mark Bahnisch, "Why We Should Be More Canadian," *Larvatus Prodeo*, 31 Oct. 2005, http://larvatusprodeo.redrag.net/2005/10/31/why-we-should-be-more-canadian/ (accessed 16 Nov. 2005).

26. See http://www.progressivebloggers.ca/.

27. See http://www.bloggingtories.ca/.

28. See http://larvatusprodeo.redrag.net/.

29. See http://backpagesblog.com/.

30. See http://www.roadtosurfdom.com/.

31. See http://johnquiggin.com/.

32. See http://troppoarmadillo.ubersportingpundit.com/.

33. See http://www.badanalysis.com/catallaxy/.

34. See http://timblair.net/ee/index.php.

35. However, there are some Labor-supporting blogs—such as *Redrag* at http://www.redrag.net/, *Wsacaucus* at http://www.wsacaucus.org/, and *Stoush.net* at http://stoush.net/; and Australian Democrats Senator Andrew Bartlett runs a prominent blog at http://www.andrewbartlett.com/blog/.

36. See the links included in Bahnisch, "Why We Should Be More Canadian."

37. Christopher Sheil, "Following the Proud Highway," *Back Pages*, 13 Aug. 2005, http://backpagesblog.com/weblog/archives/000728.html (accessed 16 Nov. 2005).

38. Mark Bahnisch, "Political Blogs versus Big Media? It's the Wrong Question to Ask," *On Line Opinion*, 22 Apr. 2005, http://www.onlineopinion.com.au/view.asp?article=3376 (accessed 16 Nov. 2005).

39. Christopher Sheil, "Blogging & Politics," *Back Pages*, 22 Aug. 2005, http://backpagesblog.com/weblog/archives/000729.html (accessed 16 Nov. 2005).

40. Bahnisch, "Why We Should Be More Canadian."

41. In a Derridean sense—see Jacques Derrida, *Limited Inc.: Abc* (Baltimore: Johns Hopkins UP, 1977).

42. George Lakoff, *Don't Think of an Elephant! Know Your Values and Frame the Debate* (Carlton North, Vic.: Scribe, 2005).

43. See the study by *Crooked Timber* bloggers discussed in Eszter Hargittai, "Cross-Ideological Conversations among Bloggers," *Crooked Timber*, 25 May 2005, http://crookedtimber.org/ 2005/05/25/cross-ideological-conversations-among-bloggers/#more-3358 (accessed 16 Nov. 2005).

44. See Max Weber's classic *Politics as a Vocation* (1919)—online at http://www2.pfeiffer.edu/~lridener/DSS/Weber/polvoc.html (accessed 16 Nov. 2005).

45. Chantal Mouffe, *On the Political* (London: Routledge, 2005).

46. As this post on the decontextualized American reception of an earlier Australian post demonstrates: Kate, "Women Want You to Shut Up (No, Not You, or You Either, Just That Guy Who Can't Find the Women Bloggers, Okay?)," *Larvatus Prodeo*, 24 Aug. 2005, http://larvatusprodeo.redrag.net/2005/08/24/women-want-you-to-shut-up/ (accessed 16 Nov. 2005).

47. Dunlop, "If You Build It They Will Come."

48. Ken Parish, "Bartlett Bleak on Blogging," *Troppo Armadillo*, 6 Apr. 2005, http://troppoarmadillo.ubersportingpundit.com/archives/008878.html (accessed 16 Nov. 2005).

CHAPTER FOURTEEN

Posting with Passion:
Blogs and the Politics of Gender

Melissa Gregg

A woman does not want the truth; what is truth to women? From the beginning, nothing has been more alien, repugnant, and hostile to woman than the truth—her great art is the lie, her highest concern is mere appearance and beauty.

—Texas Hold-em trackback spam[1] (2005)

I decided that if he noticed my notebooks, I'd say that I was writing in my diary. A safe girl-thing to be doing. Not that I had anything to hide from Detective Malloy. I just know from experience that trying to explain what it is that I write, what it is that interests me, makes me sound a little foolish, a little ineffective.

—Susanna Moore (1996, pp. 15–16)

One of the more consistent debates to have accompanied blogging's growing popularity has been the degree to which gender differences have manifested in the use of this emerging medium. Whether it is among blogging observers or participants, such discussions take qualitative and quantitative forms: on the one hand, the content of women's—as opposed to men's—blogs is seen by many to be significant, while it is the prevailing perception that the most influential bloggers are men—despite statistics showing the actual number of bloggers to be relatively even across gender lines—that remains troubling for many feminists. Each of these perspectives is in some way concerned with degrees of *recognition*: the content of women's blogs is perceived by some to be less noteworthy than men's by virtue of their often domestic and personal sphere of reference, whereas men's blogs are often seen to be more engaged in political debate, especially when the notion of what counts *as* political remains undefined. Generalizations therefore serve to confirm ingrained notions as to the proper participants in, and issues appropriate for, the public sphere.

To the extent that these regular and typically inflammatory discussions are becoming almost ritualistic, a number of feminist bloggers now take as their primary function the role of promoting and highlighting women's issues as they appear online and in the blog world in particular. It is worth acknowledging from the outset that this feminist discourse tends to be noticeably white, college-educated, and U.S.-centric. A recently developed blog, *Women's Autonomy and Sexual Sovereignty Movements*, exemplifies that key debates tend to hinge on legal matters such as the right to abortion and contraception.[2] The blog's manifesto begins by claiming that "Women are free citizens of the United States," which raises the question, Where do non-U.S. women gain recognition for their quite different and culturally specific concerns in the blogosphere? Further, while significant research has been done on the use of Websites and online communities as a means of asserting queer identities in non-Anglophone contexts,[3] there is little research currently available that looks at blogs specifically in terms of such self-expression. That being said, specific site hosts, filters, and Web-rings draw together blogs produced by and for women as part of a wider movement to lessen the offense women bloggers have understandably taken at the well-intended if naïve question, "Where Are All the Women Bloggers?" Group blogs such as *Misbehaving.net* tackle the misconception that women are somehow not as interested in technology as men, or that they have not been involved in its development, while blogs including *Ms.Musings*, *Feministe*, and *Feministing* review daily news from a feminist perspective. *The Progressive Women Bloggers Ring* draws together over two hundred self-identified feminist blogs online, while *Feminist Blogs* compiles content from affiliated members to offer "independent alternatives to the malestream media."[4] The *Blog Sisters* site, to take a further example, includes the subtitle "where men can link, but they can't touch"—a reference to the suspicion held by many women bloggers that their writing is only of interest to men when it describes sexual encounters or fantasies, or when the blogger includes pictures of herself for readers to gauge her attractiveness.[5]

These measures to increase the kinds of visibility granted women bloggers, as well as appreciation for the depth and variety in their blog content, are surprising given that women have been involved in developing some of the most popular blogging software around.[6] It is also puzzling given that the early hype surrounding the Internet's utopian possibilities put such emphasis on the liberating and playful opportunities it offered those looking to escape the confines of gender identity. Indeed, it is strange in this context just how much attention has been given to cases of successful gender fraud by award-winning bloggers, most notably "Bizgirl: International Librarian of Mystery," who played on the erotic cliché of the sexy librarian and was eventually revealed to be a man.[7] Still, the existence of Web-rings that promote themselves as "gender-free zones" testifies that the "virtual" world is vital for many individuals seeking refuge from the difficulties and dangers of non-normative gender identification offline.[8] Blogs offer a safe and fairly anonymous forum in which issues of concern and potential threat can be raised and discussed without fear. This is an important function whether in conservative local contexts or a wider political climate increasingly reflecting the values of competing religious ideologies.

Sugar and Spice and Everything Nice: Women's Blogs

Within a wider discourse of cyber-topianism, blogs have been celebrated for their capacity to reflect experiences that have been trivialized, denigrated, or ignored in the past, particularly the views of women and younger members of society. In this way, blogs are helpful for breaking the isolation many women have felt when faced with ongoing societal expectations that they are the "natural" partner to stay at home, raise children, and attend to housekeeping tasks. Blogs have been used to recount the joys encountered in a typical day to show readers (and perhaps sometimes partners) the richness of day-to-day parenting.[9] They are also used to reflect upon the responsibility of being a parent, acknowledging fears, seeking advice, and gaining encouragement from other readers. Stay-at-home mothers who blog reveal for a wide audience the actual conditions women face in the home, granting overdue recognition for the demanding and time-consuming nature of childrearing and housework. It is no coincidence that one of the strongest retorts to the question "Where Are All the Women Bloggers?" has been that women simply have less time to blog because of the unequal distribution of labor between genders. If anything, gender inequality in blogging demonstrates the reality of the "second shift."[10]

Aside from immediate family concerns, women are also perceived as spending more time than men blogging about pets, hobbies, and other domestic concerns. Knitting, cooking, and cats are the three topics typically associated with women's blogs, leading to blogger in-jokes like "Where Are All the Male Knitting Bloggers?" While these perceptions are little more than generalizations, it is nonetheless the case that a local, domestic sphere of reference is a typical response to a subordinate position within wider structures of society.[11] A domestic focus fits within a history of interests—including reading and journaling as activities—that women have developed in response to their systematic exclusion from more public forms of participation, long before information technology has become such a prestigious form of technological competence. Indeed, feminist commentators are quick to point out that when a technology is used mostly by women (the telephone or the washing machine, for instance), its value within society tends to lessen.[12]

In this still early stage of blogging commentary, critics appear to have stalled at the point of celebrating the opportunity blogs provide for expressing women's unique preoccupations and pastimes. And while this celebration is necessary, it does little to explain *why* the activities women write about on blogs might be considered more or less feminine practices. In a Western cultural context that has long separated the public (the official and sanctioned) from the realm of the private, blogging research must begin to grapple with the social and historical factors leading to women's relegation to the inherently inferior sphere of the latter. An uncritical celebration of so-called feminine practices will only perpetuate the assumption that men are active agents leaving the home *to work* while women merely tend to the home's reproduction, as if this were not also an exercise in labor.[13]

Stereotypes, Subcultures, and Support: Journals and Gender

Aside from the way in which certain *content* on blogs is seen to display gender differences, technical features of blogging software are also argued to demonstrate gendered forms of behavior. The adage "blogs are for boys, journals are for girls" summarized early observations that online diaries such as *LiveJournal* (LJ) served as natural extensions of the highly personal and intimate practice of teenage girls keeping a diary. As Danah Boyd notes, the distinction between blogs as "amateur journalism" and journals as "public diarying" creates a dichotomy and therefore cannot adequately reflect the complexity of either form: "in terms of identification, there is often a split. Most people who use LiveJournal talk about their 'LJ,' *not* their 'blog.'"[14] With LJ in particular, the emphasis on interaction, conversation, and communities of friends enabled by the software has been argued to facilitate girls' "naturally" chatty disposition. LJ reflects a different relation to readers than blogs tend to allow, because a journal page is often simply a means of entering and keeping tabs on a community of friends. In contrast to this, blogs appear to be more of an opportunity to espouse one's singular opinion.

In the sense that LJ acts as a social device for its users, it marks a new step in the practice of journaling. It allows for the public expression of thoughts that once remained inwardly directed. Blogging software here creates a form of solidarity and community-in-isolation that was rarely possible with personal diarying. As Boyd argues, this is particularly important for vulnerable members of society and those facing difficult and lonely periods in their lives (adolescence is only one example of this): "LiveJournal supports some of the most at-risk individuals, the most explicit subcultures" in contemporary culture, according to Boyd, and therefore acts as something of a social service within the wider community.

The significant support that journaling offers to those displaying nonnormative gender behavior has been demonstrated in the recent case of *MySpace* user "Zach." Zach's online journal documented his parents' decision to send him to a Christian camp run by the group "Love in Action" in an effort to reform his professed homosexuality.[15] After being picked up by a number of high-traffic blogs, Zach's journal was inundated with hundreds of statements of support from readers alerted to his situation. While the long-term effects of this case will only be known later, it is a topical example of the manner in which online communities offer solace for those whose identity proves to be questionable and even subject to discipline within the immediate geographical and social location of the blogger.

Zach's case serves to show the crucial role online journals can play in situations where role models for non-normative behavior are few. Yet within blogging culture, the phenomenon of "LiveJournal bashing"—mocking the interests of online journal writers[16]—arises from the assumption that the personal chat of young people is trivial in comparison to the weighty political content discussed on pundit-style blogs. An implicit power dynamic enables this distinction, which deems some issues to be trivial while others are more significant. As the opening quotation for this chapter indicates, such a distinction also follows a long tradition of philosophical thought that places women's culture in the domain of emotion and affect as opposed to the rationality and reason of which men are capable. The debate about gender and blogging has therefore suffered from a lack of clarity in three main areas: what counts as a blog, what counts as an online journal, and what counts as political. Mainstream perceptions of blogs, when they exist at all, currently gravitate between the stereotypes that blogs are online journals written by young women about their personal lives or that they are the territory of older men seeking status as political pundits or provocateurs. As other researchers have argued, however, these perceptions create a hierarchy whereby the group or pundit blog—sometimes called the "filter" blog—is the authentic form against which other styles of blogging must be judged:

> by privileging filter blogs and thereby implicitly evaluating the activities of adult males as more interesting, important and/or newsworthy than those of other blog authors, public discourses about weblogs marginalize the activities of women and teen bloggers, thereby indirectly reproducing societal sexism and ageism, and misrepresenting the fundamental nature of the weblog phenomenon.[17]

What these debates also typically avoid is any significant debate about what makes a topic "political," "newsworthy," or "important" in the first place, and the history of gender bias within the bastions of knowledge production (such as educational, media, or political institutions), which have set the terms of public debate for so long. In blogs, two forms of "mastery" combine: that of technological competence and that of authoritative speech. Both are forms of mastery gained by a past economy of exclusion, which is to say that they rely on a familiarity with, and access to, various forms of literacy that for the most part have only been available to white men until very recently. If distinctions do exist between women's and men's interests on blogs, this may still be a reflection of previous restrictions placed on education on the one hand and participation in the public sphere (as a corollary) on the other. Against this history it is only understandable that there is a lag in priorities and some variation in interests; the opportunity blogging offers is that this kind of unequal access, and strict division between the public and private sphere, remains a thing of the past.

It is also against this history that we might question the logic behind the following statement, representative of many canvassed during blogging's gender wars:

> women on the whole are less interested in politics than men, therefore less women create blogs, thus the female talent pool in the blogosphere is smaller than the male pool, which leads to the dearth of "A-List" female bloggers. ... In other words, there aren't as many really successful female bloggers because percentage wise, there aren't as many women who are interested in doing political blogging. It's just that simple.[18]

While there is much to debate in this passage, the logic behind this position is that politics itself is static, that what counts as political does not need to be debated because it is self-evident. Not only is this truly worrying for the prospect of any kind of effective political change or agitation, as it assumes an unchanging list of priorities for political debate, but it refuses any mention of the second-wave feminist movements of the 1960s and 1970s that established personal issues *as* political. Feminism has shown that there is no easy separation between individual experience and political perspective, and to demonstrate this point a little further, we need only take note of a subsequent post from this same pundit, outlining how many blogs he reads per day:

> on a slow day, I may only hit 40–60 blogs and websites, but typically I hit somewhere in the neighborhood of 75–100 blogs and websites a day Sun–Fri. Here are the blogs that I have bookmarked as "daily reads," which in practice, means that I hit them 3–5+ times per week. Hopefully, you'll find some new reads....[19]

The amount of time such a routine would demand indicates the kinds of conditions typically required of those seeking to participate in "political" debate on blogs. It requires predictable priorities, a distinct method, and above all a lack of interruption from offline demands. This helps to illustrate that time influences women's participation in political punditry blogs in two senses. First, a continuing unequal division of labor within the home affects women's sheer ability to keep track of debates among bloggers and the amount of time required to write regularly enough to sustain a wide readership. Second, women continue to fight the consequences of the amount of time men have enjoyed occupying positions of power in the public arena, time spent establishing their outlook as rational, right, accomplished, and authoritative.[20]

Alternative Spaces, Different Outlooks

My own blog, *Home Cooked Theory*, records the experiences of being an academic from the perspective of a young woman who is in her first full-time job.[21] I write about everyday issues having to do with my work, but I also use the space as a way to think out loud about current projects I'm involved in. This is part of an attempt to demystify some of the isolation and secrecy of academic practice, to hold my research accountable to the public that supports it. The title of the blog indicates its intention, which is to question the "traditional" locations for intellectual and political debate. Shifting between more personal matters and the political issues at the heart of my research concerns, the blog avoids any fixation on matters deemed important within the public sphere. When referred to, such material tends to be tempered by, or situated within, the blog's professed interests.

Since starting my blog I've found that my writing tends to elicit more responses from men than women, which worried me enough that I raised it for debate on the blog itself. Commenting on her hesitation to engage in blogging discourse, one reader explained:

> the "thrust and parry" approach that many bloggers adopt is, for me, exhausting and unproductive—unless you are training to be a professional debater. This will undoubtedly prove to be a controversial observation, but there seems to be a fair bit of bravado informing these textual performances.

While for many such a position could be read as a reference to gender differences in blogging encounters and the performative style demanded by the technology, the reader goes on to add:

> I suspect this may also have something to do with the structural and temporal limitations of an electronic "exchange" which doesn't function as a fluid real-time conversation between participants, but as a series of (at times inter-connected) scriptural fragments.[22]

This example is useful to highlight the idea that gender differences can be ascribed to a lot of the behavior that takes place on blogs, just as they can in "real life," but to fully appreciate how the medium functions requires a multi-faceted approach that doesn't rely too heavily on one interpretive lens. In assessing the uses of blogs, it's dangerous to start with the assumption that new media will necessarily replicate hierarchies of access and recognition that exist elsewhere, or worse, that technologies themselves are inherently gendered. But nor are we naïve. The performance cues of particular individuals who blog can certainly be read within a history of gendered practices and privileges, and it shouldn't be such a laborious and necessary task to point out those recurring instances where gender does (or conversely, fails to) become a factor.

At this early stage of reflection, as blogs are called into meaningfulness through commentary, it is important that they aren't trapped within vocabularies and categories that fail to reflect their unique potential.[23] During this exciting time we should remain wary of any factors that may affect our ability to recognize a new medium or practice for and of itself. Indeed, perhaps the most fitting way to conclude a chapter on blogging and gender is to point out that it is partnerships between women and men that have enabled so many of the crucial developments in blogging so far. As I've mentioned, Meg Hourihan and Evan Williams formed the company that developed *Blogger*, but Mena and Ben Trott co-founded Six Apart, the company behind Movable Type, *TypePad*, and now *LiveJournal*. (Six Apart is named after the fact that the married couple were born six days apart.) *Flickr*, the photo-sharing Website servicing many blog platforms, was developed by another couple, Caterina Fake and Stewart Butterfield. This unfolding history tends to indicate that sharing and valuing our distinct perspectives between and across genders puts us in the best position to fully realize blogging's still unknown potential.

NOTES

1. Texas Hold-em is the name of one of the most prolific spammers to have targetted Web-logs, employing a difficult-to-control technique that operates by "trackback" pings, a reference function in many Weblogs. The original source for the quotation is Friedrich Nietzsche, *Beyond Good and Evil*, first published in 1886, and I use it here to indicate the long history of gender stereotypes affecting Western thought and against which blogging commentary must be understood.

2. http://the-goddess.org/wam.

3. Mark McLelland, "Private Acts/Public Spaces: Cruising for Gay Sex on the Japanese Internet," in Nanette Gottlieb and Mark McLelland, eds., *Japanese Cybercultures* (London: Routledge, 2003), pp. 141–55; Mark McLelland, "Live Life More Selfishly: An On-Line Gay Advice Column in Japan," *Continuum* 15.1 (2001), pp. 103–16.

4. http://msmusings.net/; http://feministe.us/blog; http://www.feministblogs.org/; http://feministing.com/; http://www.feministblogs.org/.

5. See http://blogsisters.blogspot.com/. Clancy Ratliff, a Ph.D. candidate in the Department of Rhetoric at the University of Minnesota, has compiled an extensive list of blogging and gender debates on her blog, *Culture Cat*, http://culturecat.net/node/637. Ratliff's thesis promises to be the first in-depth study of the way blogging can be seen to challenge gendered notions of political discourse and the public sphere.

6. Meg Hourihan, blogger and contributor to one of the first books dealing with blogging (Haughey *et al.*, 2002) helped found Pyra, the company that created Blogger. She recounts the story of Blogger's development at IT Conversations, available at http://www.itconversations.com/shows/detail541.html.

7. Bizgirl's blog remains at http://bizgirl.blogspot.com/.

8. See "Gender Free Zone" at http://www.steeltoed.com/genderfree/index01.html and "Love Knows No Gender" at http://m.webring.com/hub?ring=bipeoplewholove.

9. Winner of the Best Australian Personal Blog in the 2005 Australian Blog Awards, *She Sells Sanctuary* recently featured a series of posts listing the many things the writer loves about her son, http://she-sells-sanctuary.blogspot.com/2005/06/counting-ways.html. Yet this blog, written by single mother Gianna, also responds to news stories of the day, particularly those that impinge on her everyday life and identity as Australian citizen. Gianna's blog is strong evidence of the manner in which women are often interested in politics as it is conventionally defined, yet the particular demands of their situation often mean that such issues are contextualized within more pressing daily matters.

10. Arlie Hochschild with Anne Machung, *The Second Shift: Working Parents and the Revolution at Home* (New York: Viking, 1989).

11. Richard Hoggart, *The Uses of Literacy* (Harmondsworth: Penguin, 1958).

12. Dale Spender, *Nattering on the Net: Women, Power and Cyberspace* (North Melbourne: Spinifex, 1995).

13. Meaghan Morris, "At Henry Parkes Motel," in John Frow and Meaghan Morris, eds., *Australian Cultural Studies: A Reader* (St. Leonards: Allen & Unwin, 1993).

14. Danah Boyd, "Turmoil in Blogland," *Salon*, http://www.salon.com/tech/feature/2005/01/08/livejournal/index.html (accessed 19 July 2005).

15. At the time of writing, Zach's page was available at: http://profile.myspace.com/index.cfm?fuseaction=user.viewprofile&friendID=7428306&Mytoken=20050718175635.

16. Mathieu O'Neil, "Weblogs and Authority," paper presented at *Blogtalk Sydney*, http://incsub.org/blogtalk/?page_id=107 (accessed 29 July 2005).

17. S.C. Herring, I. Kouper, L.A. Scheidt, and E.L. Wright, "Women and Children Last: The Discursive Construction of Weblogs," in L.J. Gurak, S. Antonijevic, L. Johnson, C. Ratliff, and J. Reyman, eds., *Into the Blogosphere: Rhetoric, Community, and Culture of Weblogs*, 2004, http://blog.lib.umn.edu/blogosphere/women_and_children.html (accessed 29 July 2005).

18. John Hawkins, "Why There's a Dearth of A-List Female Bloggers," *Right Wing News*, http://www.rightwingnews.com/archives/week_2005_02_20.PHP#003488 (accessed 19 July 2005).

19. John Hawkins, "The Blogs I Read Daily," *Right Wing News*, http://www.rightwingnews.com/archives/week_2005_07_17.PHP#004128 (accessed 19 July 2005).

20. For women interested in joining the ranks of A-List punditry or filter blogs, a key complaint has been that men rarely link to women bloggers, and that women comprise smaller numbers among both contributors and blogrolls on filter and pundit-style blogs. Many male bloggers have since acknowledged this as the case, yet such observations fail to acknowledge that links from another blogger may not signify respect or esteem—that linking can have negative as well as positive intent. It also betrays the assumption that blogging is best understood within a framework of celebrity, when for many bloggers, women and men, it is the quality rather than the quantity of readership that makes writing online so rewarding.

21. See http://hypertext.rmit.edu.au/~gregg.

22. "Kirsten," Comment on "Call for Proposals," *Home Cooked Theory*, April 2005, http://hypertext.rmit.edu.au/~gregg/archives/2005/04/29/call-for-proposals/ (accessed 29 July 2005).

23. Kris Cohen describes the challenge such an open approach poses for critical commentary in "A Welcome for Blogs," *Continuum Special Issue: Counter-Heroics and Counter-Professionalism in Cultural Studies*, 20.2 (June 2006).

CHAPTER FIFTEEN

Blogging Disability:
The Interface between New Cultural
Movements and Internet Technology

Gerard Goggin & Tim Noonan

S ince their inception in the late 1990s, there has been much excitement
about blogs and their possibilities. Little noticed by mainstream bloggers
and Internet, communications, and media studies at large, people with
disabilities have been among those flourishing in the fertile world of the blogosphere.

In this chapter we wish to introduce the reader to this important aspect of
blogging, but before we delve into blogging disability, we wish to establish
what disability signifies for us, why we consider it as a new cultural movement,
and what the relations are between disability and media such as the Internet.

Disability and the Internet

For the past century at least, disability has been typically thought of as a
deficiency or terrible loss. The charitable approach to disability is still very
powerful, positioning people with disabilities as individuals who should be
pitied and assisted by magnanimous, selfless
benefactors. The biomedical model of disability
revolves around an image of people with disabilities
as having bodies that need to be cured, or at least
treated and rehabilitated, by the enlightened efforts
of medical and health professionals. Our approach
to disability challenges these dominant
understandings.

> **Disability:**
> the social and cultural
> shaping of impairment;
> disability is shaped by
> power relations (cf. gen-
> der, race, sex, class).

Informed by the movement of people with disabilities that has arisen since
the 1970s, we consider disability as a sociopolitical process. In doing so, we

draw upon the ideas and work of what has been variously called "new" or "critical" disability studies. The central tenet of this critical disability studies is to radically call into question the fixed idea of disability and its location in "deviant," disabled individuals. British disability movement activists and scholars (especially sociologists) have famously proposed a binary opposition between "impairment" and "disability." They suggest that impairment is the material, bodily dimension (the "objective" sensory condition of blindness, for instance), as opposed to disability, which is what society makes of that impairment.

Proponents of the social model point out that many of the difficulties and barriers people face are people-made and socially constructed. If information technology is not designed with the desires and capabilities of people with disabilities in mind, then it can be disabling. Disability does not reside with, nor is the fault of, the person with disability; it is something brought about by an oppressive, inequitable, and careless set of social relations. This is documented in a study entitled *Digital Disability: The Social Construction of Disability in New Media*, which looks at how, time and time again, much vaunted "new" technology is needlessly inaccessible to people with disabilities.[1] The social model has been critiqued by a range of theorists who have pointed out the shifting and complex relationships between body and society, matter and idea, nature and culture, that are not well explained by a fastidious adherence to this disability/impairment couplet.[2]

These new approaches to disability have been used to better understand and explain media. Studies have shown the way that representation of disability in media is an important part of the discursive shaping of disability in society and have critiqued the lack of participation of people with disabilities in media organizations and the neglect of people with disabilities as media audiences.[3] The Internet has offered new possibilities for people with disabilities to access content and communications that hitherto have been inaccessible to them in traditional print or audiovisual media, and also, as for other users, has provided new spaces and structures for becoming media and cultural producers. The use of the Internet by people with disabilities has not been given the theoretical or empirical attention it deserves,[4] but some useful studies are now available.[5]

Accessibility, Participation, and the Social Shaping of Disability in Blogging

It can be said that for every new technology that is developed, there will be people with disabilities who will wish to access, use, and participate in the activities associated with that technology. And blogging, of course, is no exception. There are three main areas where blogs can present accessibility problems for people with disability: establishment of a blogging account, maintaining a blog, and "reading" blogs.

Before a person can start posting to their blog, they need to create or establish a blogging account. In some instances—*Blogger* is currently a case in point—the maintaining and accessing of blogs presents minor access challenges, but (quite ironically) the registration process can be almost completely impossible for many disabled users to complete without assistance from another. *Blogger* uses a "captcha" (or "completely automated public Turing test") to tell computers and humans apart.[6] A captcha is a concept developed and designed to make it very difficult for automated computer "bots" to establish all manner of user accounts on Internet services. (Ironically, the most reliable version of this system, which entails a user reading an image of distorted text and typing it in, is called "Gimpy.") While this helps cut down spam, it also creates almost insurmountable road blocks for some people with disability, such as many Blind people.[7]

Once a blog has been set up, the accessibility challenge for many users is reduced. For instance, depending on the blogging solution selected, screen readers, screen magnifiers, or Braille display output are generally achievable. Some blogging solutions are more accessible than others, and some offer a variety of interfaces for posting, such as from plain text email. Accordingly, difficulties due to poor design are surprisingly common. For example, even to comment on another's blog requires captcha input with *Blogger*, posing real barriers for Blind people to creatively interact with such blogs. As blogs for the most part are based on modification and innovation of Web technologies, if they conform closely to the Web Content Accessibility Guidelines of the World Wide Web Consortium (W3C), they will be easier for a range of people with differing disabilities to contribute to, and to read.[8]

> **Accessibility:** the potential for a technology to be used by a wide range of users (or even *all* users) without barriers being placed in their way through poor or discriminatory design (or the social shaping of the technology).

There are two areas we think are important when considering participation for bloggers with disabilities. First, there are the issues that arise from the unique practices entailed in blogging that differ from other Internet media production and consumption modes, and which may present particular issues. For instance, the closely interlinked nature of the blogosphere, with blogs and blog posts ineluctably linking with other blogs, underscores the need for all programs, technologies, and protocols to be designed with affordances for accessibility. Second, there are considerable challenges already faced by the information and communications technology industries in making accessible their own products and ensuring that not only are these interoperable and

achieve interconnectivity across networks, but also that there is end-to-end accessibility for users. Accessibility issues have been most prominently and successfully addressed with respect to the Web by the W3C Web Accessibility
Initiative. However, many standards and industry bodies and companies, regu-
lators, and user groups are grappling with the many other technologies at play
in constituting the Internet, not least in multimedia software, peer-to-peer
technologies, and the widespread deployment of new, converged communications such as Voice and Video over Internet protocol, and also the unfolding
area of mobile Internet (especially with third-generation mobiles). The example
of sign language blogs is instructive here: meaningful accessibility requires sufficient two-way bandwidth (at least ISDN or, better still, broadband) in order
for sign language to be captured, read, and responded to. Third, there is a real
need to move beyond talking about accessibility to considering the deeper issues in participation, cultural consumption and citizenship of people with disabilities. Here, especially, there is a real need to critique the all-too-easy idea
(but widely held) that the W3C guidelines or other standards or even "universal design" practices are all that is required.[9]

Crip Blogs: Mapping Blogs by People with Disabilities

By late 2005, there was already a bewildering array of blogs by people with disabilities. However, we were unable to find any studies of disabled blogging.
One possible way of approaching blogging by people with disabilities is to consider how they figure in demographics of blog authors. While there have been
some quantitative surveys, including demographics of blog authors by age and
gender[10] and some helpful discussions of gender and blogs,[11] there has been no
such discussion of how many people with disabilities are blogging. This question could be approached, of course, by first considering other surveys and statistics on people with disabilities as part of the overall Internet user
population—of which there appear to be very few anyway—but to extrapolate
from such would not only be potentially misleading, but could also tell little
about the participation and distinctive patterns of use of disabled bloggers.
Pursuing quantitative studies to map the disabled blogosphere poses its own
methodological problems (given the dynamic, complex, and often invisible nature of disability) and is outside our scope here. Instead, we will briefly discuss
some disabled blogs, often cited in blogrolls, search engines, and other online
and mainstream media, as well as blogs that recurred in our conversations with
bloggers.

The first type of blog we would point to is that explicitly articulated from a
disability activism or disability studies perspective. A leading example is *Gimp
Parade*, a blogger who introduces herself as a "thirtysomething disabled feminist. Overeducated, underemployed": "A website with personal reviews of

books and movies from a disability culture perspective."[12] *The Gimp Parade* moves lightly between the political, academic, and personal, in a way that is very much representative of disability studies discourse. *General Observations Using Disability Studies* also starts from this premise.[13] There are now also a growing number of institutional or organizational blogs maintained by academic programs, such as *Disability Studies* at Temple University,[14] as well as a variety of academic and research blogs by scholars for whom disability is one of a number of interests, such as that of Michael Bérube[15] or Karen Nakamura's *Photoethnography.Com.*[16] Another academic blog, but especially devoted to Internet and blog studies, is Katherine Mancuso's *museumfreak.*[17] Among other interests, Mancuso is studying subcultural groups on *LiveJournal* as a community of practice. (*LiveJournal* is a hugely popular blogging site, under study by a number of Internet scholars, and Mancuso points to the prevalence of disability blogs hosted there.)

Another broad sort of disabled blog is the burgeoning number of activist and information blogs, often with a focus on a particular perspective or topic.[18] By far the largest category of self-identified blogging by people with disabilities is of a diaristic or journal nature. There are many blogs whose prime purpose is to share someone's experience of life, including—as a part of that person's life—thoughts and recountings of impairment and disability, and to share this with whatever large or, more realistically, small (or micro) audience might be interested.[19]

In thinking about how to map the wonderfully chimerical disabled blogosphere, we would point to the wide range of communities for whom the Internet, digital technologies, and media are an important part of the constitution of their identities and cultural traffic. For instance, at the intersection of disability and health, and the redrawing of "health" as disability (note, for instance, the many studies on health and the Internet compared to the small number on disability and the Internet), there are many bloggers active in the "mental health Internet community" (as Natasha Kraus has pointed out to us). Indeed, the question of disabled identity is a complex and much-debated one, especially with relation to the Internet, where "passing" as non-disabled is potentially possible in new ways. The *Blind Insight* blog, for example, demonstrates the public journaling of thoughts and attitudes about having low vision, but still being able to "pass" as sighted.[20]

One of the other interesting things that some disabled blogs are doing is to move beyond the medical versus social model divide of disability that has deeply shaped the constitution of disability, in exploring a variety of dimensions of life for people with chronic conditions who depart from some of the stereotypes and "rules" with regard to disability. Indeed, blogs may be seen to provide alternative narrations that are not necessarily in accordance with the dominant paradigms, including those proposed by social model theorists.

Emerging practices in blogs suggest that it is not so easy to draw the distinction between health and disability, especially on the Internet, and in fact many people with chronic conditions cannot (or choose not to) do so.[21]

As well as focusing on people with disabilities, another way to approach our topic is to interrogate how disability is implicated in the "discursive construction" of Weblogs. An initial hypothesis here could be how blogging and the blogosphere are shaped by what appear to be naturally given, "normal" concepts of blogging practice and authorship, yet how these code particular disablist accounts of blogging. In a suggestive study of female and teen bloggers, Herring *et al.* argue that the relative absence of women and teens from public discourses about Weblogs goes to what sorts of blogs are privileged, namely "filter" blogs, which they view as largely produced by adult males and consistent, it might be added, with the valorization of news, current affairs, and "serious" commentary in accounts of media.[22] The implication of this argument is that a more comprehensive mapping of blogging is needed to ensure that it captures, and gives due weight, to all blogging activities, especially online journals. (This observation was one impetus for this very book, of course.)

Having noted some of the features, but also quandaries, in thinking about the disabled blogosphere, we now wish to turn to an area where disability and blogging has contributed important innovations to blogging practice, namely multimedia and multisensory blogging.

Blind Blogging

> **Remediation:**
> the representation of one medium in another, but particularly the way that new media draws upon and reworks old media.

Blogging still remains much dominated by textual practices and the remediations of the histories of these forms. However, the Internet, and blogging in particular, offer new modes for people with disabilities to author, communicate, consume, and exchange in their preferred medium or media. Aficionados would be familiar with the growing popularity of photoblogging, video-blogging (vlogging), moblogging, and audioblogging, and their affiliated technologies and practices (the sharing of images through social software and photo sharing sites such as *Flickr*, for instance). Less well known, but more significant in many ways, are the practices used by some Deaf bloggers (with sign language blogging enabled by video) and the case we now wish to briefly discuss, audio and other practices used by Blind bloggers.

While blogs have attracted reasonable attention and uptake by people who are Blind, podcasts have been taken up by Blind people even more resoundingly, both in creation and consumption. Podcasts intrinsically exemplify the oral tradition, something that many Blind people feel very comfortable with, and feel is natural to them. This is demonstrated by the fact that there are at least two key resources highlighting Blind-related podcasts,[23] yet by contrast none that we are aware of that focus on Blind bloggers. Further, many blogs of Blind people either incorporate links to audio ramblings, hard-hitting journalistic inquiry, or sharing of technology barriers and solutions of interest to people with disability.

> **Podcasting:**
> a way to distribute audio and video programs through the Internet; users subscribe to "feeds" pointing to audio files that are downloaded to their portable player (iPod or MP3) for them to play when they wish.

A good example is *The Mosen Explosion*,[24] maintained by Jonathan Mosen, well known for his extensive work in terrestrial broadcast radio and, in recent years, Internet radio. An internationally famous personality in his own right, Jonathan's blog cuts across diverse dimensions of his life, family (as a Blind father of children without disabilities), accessibility (recognized as an expert interviewer and, more recently, technology product marketing manager), radio (operating a station), Blind politics journalism, and as a huge fan of the Dr. Pepper beverage. His blog expresses multiple dimensions of his life, including but not primarily focusing on his blindness (an aspect of him, but clearly not defining him). Jonathan's blog contains a mix of textual and audio entries, reflecting his dual love of news and spoken commentary.

Blindspot: Marlaina by Ear is another important blog.[25] Marlaina combines her blogging with podcasts and her weekly program on American Council for the Blind radio in North America (but broadcast internationally via the Internet).[26] Jeff Bishop's *Desert Skies* blog, podcast, and Internet radio show focuses on technology developments and accessibility for Blind people.[27] Also widely cited is *Blind Chance: David Faucheux's Audio Web Log*,[28] no doubt due to his thought-provoking audio posts and eclectic interests, but perhaps also because he is a Blind librarian. All three of these blogs illustrate the nexus of technologies now involved in blogging, and the implications for information flows and communications. Previously, Internet radio and electronic mailing lists were the key vehicles for this kind of content for Blind audiences, and now blogs play a nodal role in structuring and articulating these different forms.

Disability Blogs in the Media

Many mainstream media organizations with an online presence, such as newspapers, radio, and television broadcasters, have added blogs to their Websites. Such blogs can provide richer, more elaborate, and more quickly updated breaking news from a variety of reliable as well as untested or unfiltered sources. They also offer a different sort of opportunity for reader comment and response, as well as linking to other blogs and Websites. While there are many examples of publications and media organizations that have incorporated blogs into their operations, there are very few specialist or mainstream media organizations that offer a disability-specific blog.

One of the few exceptions is a weblog established in June 2002 as a part of the British Broadcasting Corporation's *Ouch!* Website. The aim of *Ouch!* is "to reflect the lives of disabled people right here and now in the third millennium."[29] It is unclear when the *Ouch!* blog was established, but its archive carries entries dating back to April 2004, and it is multi-authored by the *Ouch!* team.[30]

The blog offers disability comment and perspective on issues in daily life and popular culture familiar to many people with disability. It is especially acute on critique of media and disability, discussing debates about non-disabled actors playing disabled roles in movies, or disablist headlines or reportage in print media, representation of people with disabilities in television ("Is Channel Four finally going to put a disabled person into the *Big Brother* house?"), disability culture icons (Andy and Lou in the British comedy series *Little Britain*, or the satirical cartoon *Quads*), and publicizing disability film festivals. The *Ouch!* blog has a particular interest in disability and new media, often covering issues in Web accessibility, but also seeks to encourage disabled blogs.

It is useful to contrast the *Ouch!* approach to blogging, and its blogroll, with that of the well-known U.S. *Ragged Edge* magazine. While *Ouch!* seeks to provide a forum for disability perspectives on current affairs, by virtue of its location as part of the BBC it also seeks to educate and inform both those who identify as non-disabled and those who do not of disability culture. Its ethos, stance, and address are all very much informed by the values and charter of the British public service broadcaster. *The Ragged Edge*, on the other hand, is very much a media institution set in and informed by the disability movement. Rather than feature an official blog, *The Ragged Edge* has a linked and closely related blog called *Edge-Centric: Ranting and Musings from Your* Ragged Edge *Online Editor* (journalist and author Mary Johnson). Entries from the blog are selectively featured among other news on the home page of the *Ragged Edge* Website. *Edge-Centric* has a strong and at times quite powerful disability critique, very much a voice from, of, and centered in, the North American disability movement and disability studies community. *The Ragged Edge* expresses what is fast becoming a commonplace among people with disabilities online, and a view put to us by many people consulted for this chapter, namely that blogging is an important space for putting new ideas about disability: "Some of the best disability commentary around is now coming from blogs: Why not take a look for yourself? ... There are hundreds of crip blogs now—who knows? Maybe thousands!"[31]

Blogging Bodies

This chapter can only serve as an impossibly brief and partial introduction to disability and blogging and the extraordinary and ordinary excitement, innovation, and possibility that is unfolding around this. The most obvious contribution of disabled bloggers to blogging in general lies in the innovative uses of textual, visual, and audio modes of representation and communication. There are well-established, long-standing, and rich cultural and communication traditions into which such blogging may be placed: audio blogging and podcasting by Blind bloggers can be seen as elaborating and displacing earlier oral, spoken-word, and sound traditions in media artifacts such as talking-books, audiotapes, Daisy Books, CDs and DVDs, Radio for the Print Handicapped, and other forms of non-networked and network digital audio, as well as e-mail lists and Websites; sign language blogs are remediating other visual communication and media of Deaf communities, not least the importance of the face and body to signification; text blogs are drawing upon a wide range of disability cultural and subcultural genres, histories, and dispositions.

In traditional mainstream media, people with disabilities have long lamented and critiqued both their access and the ways they are customarily represented. Various forms of new media, from the advent of affordable, portable video cameras to Braille PDAs and other computers, to teletypewriters (used by Deaf people) and mobiles, but most important perhaps the Internet, have offered new resources for people with disabilities to become media producers in their own right—an old dream of the democratic potential of new media.

So, for instance, there has been an especially strong and fertile outpouring of a wide diversity of memoirs and narratives of people with disabilities over the past two decades—not least because such stories have not been widely told, yet attract substantial audiences (given the high incidence of readers, and writers, with some personal or other direct experience of disability). Blogging by people with disabilities, we would suggest, has further expanded such texts, and also, akin to blogging generally, has created new forms of reception, appropriation, and reading of such stories. Blogs sometimes are dismissed for being little more than the reflection of the normal activities of daily life, and perhaps one of the biggest and least recognized contributions of personal disability blogs is that they serendipitously illustrate greater similarities than differences between temporarily abled and disabled people. The practical and political implications of blogging with respect to disability were also dramatically illustrated in the September 2005 Hurricane Katrina in New Orleans, where blogs played an important role in the disabled as well as non-disabled communities' emergency and communicative responses to this disaster.

There are many questions that have only begun to be discussed, if at all, regarding disability and blogging, that we would note in conclusion. A recurrent theme in discussions of blogging have been divisions and distinctions that constitute the structure of the blogosphere, such as the emergence of "A-list" bloggers. If there are emerging canons, or at least mainstreams and margins, in the blogosphere, where do disabled bloggers sit? And how is disability constructed in the blogosphere, through such developments? Another theme is the incorporation of blogging into existing patterns of media ownership and control (whether in the case of blogs as extensions to well-known media outlets, or of blogging sites and software being taken over by old or new media giants). How will long-standing power relations of disability be reconfigured in such maneuvers? A third theme is understanding the distinctive cultures of blogging across different language, national, and regional communities, class, gender, race, and disability groups. We have focused here on relatively visible English-language disability blogs with many U.S., and a few British, New Zealand, and Canadian examples, but have discussed no English-language blogs from elsewhere in Oceania, Asia, or Africa, not to mention blogs in other linguistic and cultural communities well represented on the Internet (such as Spanish, Japanese, and Chinese). This will become an important direction of inquiry, especially given the dawning recognition that the Internet is very much internationalizing, and that English is well on the way to becoming a minority language here.

NOTES

1. Gerard Goggin and Christopher Newell, *Digital Disability* (Lanham, MD: Rowman & Littlefield, 2003); see also Goggin and Newell, eds., "Technology and Disability," special double issue of *Disability Studies Quarterly* 25.2 & 3 (2005), http://www.dsq-sds.org/_articles_html/2005/spring/ (accessed 30 Oct. 2005).

2. Mairean Corker and Tom Shakespeare, *Disability/Postmodernity: Embodying Disability Theory* (London: Continuum, 2002).

3. Beth Haller, "If They Limp, They Lead: News Representations and the Hierarchy of Disability Images," in *Handbook of Communication and People with Disabilities: Research and Application*, D. Braithwaite and T. Thompson, eds. (Mahwah, NJ: Lawrence Erlbaum, 2000); Gerard Goggin and Christopher Newell, "Crippling Paralympics?: Media, Disability and Olympism," *Media International Australia* 97 (2000), pp. 71–84, and "Uniting the Nation?: Disability, Stem Cells, and the Australian Media," *Disability & Society* 19.1 (2004), pp. 47–60; Karen Ross, "All Ears: Radio, Reception and Discourses of Disability," *Media, Culture and Society* 23 (1997), pp. 419–37, and "But Where's Me in It? Disability, Broadcasting and the Audience," *Media, Culture and Society* 19 (2001), pp. 669–77.

4. Goggin and Newell, *Digital Disability*.

5. Natilene Bowker and Keith Tuffin, "Disability Discourses for Online Identities," *Disability & Society* 17 (2003), pp. 327–44; Judith A. Cook *et al.*, "Information Technology Attitudes and Behaviors among Individuals with Psychiatric Disabilities Who Use the Internet: Results of a Web-Based Survey," *Disability Studies Quarterly* 25.2 (2005), http://www.dsq-sds.org/_articles_html/2005/spring/cook_etal.asp (accessed 30 Oct. 2005); Jin Huang and Guo Baorong, "Building Social Capital: A Study of the Online Disability Community," *Disability Studies Quarterly* 25.2 (2005), http://www.dsq-sds.org/_articles_html/2005/spring/huang_guo.asp (accessed 30 Oct. 2005); Becky Moss, Susie Parr, Sally Byng, and Brian Petheram, "'Pick Me Up and Not a Down Down, Up Up': How Are the Identities of People with Aphasia Represented in Aphasia, Stroke and Disability Websites?" *Disability & Society* 19 (2004), pp. 753–68; Deborah Stienstra and Lindsey Troschuk, "Engaging Citizens with Disabilities in eDemocracy," *Disability Studies Quarterly* 25.2 (2005), http://www.dsq-sds.org/_articles_html/2005/spring/stienstra_troschuk.asp (accessed 30 Oct. 2005).

6. See http://www.captcha.net/.

7. World Wide Web Consortium (W3C), "Inaccessibility of Visually-Oriented Anti-Robot Tests Problems and Alternatives," *W3C Working Draft*, 5 Nov. 2003, http://www.w3.org/TR/turingtest/ (accessed 28 Aug. 2005). In this chapter, we capitalize "Blind" and "Deaf" in accordance with understandings of these communities as cultural and linguistic minorities (cf. "French" or "American").

8. American Foundation of the Blind, "How to Make Your Blog Accessible to Blind Readers," http://www.afb.org/Section.asp?SectionID=4&TopicID=167&DocumentID=2757, and "Is Blogging Accessible to People with Vision Loss?" http://www.afb.org/Section.asp?SectionID=57&DocumentID=2753 (accessed 21 Aug. 2005).

9. For a critique of accessibility and universal design see Goggin and Newell, *Digital Disability*, pp. 147–50.

10. J. Henning, "The Blogging Iceberg—Of 4.12 Million Hosted Weblogs, Most Little Seen, Quickly Abandoned," *Perseus Development Corp. White Papers*, 2003, http://www.perseus.com/blogsurvey/thebloggingiceberg.html (accessed 20 Aug. 2005).

11. Susan Herring et al., "Women and Children Last: The Discursive Construction of Weblogs," in L.J. Gurak, S. Antonijevic, L. Johnson, C. Ratliff, and J. Reyman, eds., *Into the Blogosphere: Rhetoric, Community, and Culture of Weblogs*, 2004, http://blog.lib.umn.edu/ blogosphere/women_and_children.html (accessed 1 Aug. 2005).

12. *The Gimp Parade*, 1 June 2004, http://thegimpparade.blogspot.com/2004_06_01_ thegimpparade_archive.html (accessed 1 Aug. 2005).

13. See http://goudisabilitystudies.blogspot.com/.

14. See http://disstud.blogspot.com/.

15. See http://www.michaelberube.com/.

16. See http://www.photoethnography.com/blog/.

17. See http://www.livejournal.com/users/museumfreak/.

18. Frequently read and cited blogs include Scott Rains's *Rolling Rains Report: Precipitating Dialogue on Travel, Disability, and Universal Design* (http://www.rollingrains.com/); J. Kevin Morton's *Disability Law Blog* (http://jkm.typepad.com/); *Disability Law* (http://disabilitylaw.blogspot.com/); biotechnology activist Gregor Wolbring's *My Thoughts on Social and Scientific Issues* (http://wolbring.blogspot.com/). Further, there are blogs in other fields by people with disabilities for whom disability is an important part of their professional as well as personal identity, such as disabled writer and performer Greg Walloch's blog (http://gregwallochblog.blogspot.com/).

19. Scott Laurent's *Disability Is an Art* blog (http://www.disabilityisanart.blogspot.com/); Katja Stokley's *Brokenclay: The Art of Intermittent Disability* (http://brokenclay.org/journal/); *Becky's Journal: The Ordinary Life of a Not-So Ordinary Girl* (http://www.dalqe.com/); *Crazy Deaf Joe Blog* (http://www.crazydeafjoe.com/); *Schizophrenia Blog: My Life's Adventure* (http://www.schizophrenia.com/journey/); *A Gimp's Life* (http://journals.aol.com/ brucer5150/AGimpsLife/). Among the more widely read journal-style blogs is the witty *Lisy Babe's Blog* (http://lisybabe.blogspot.com/).

20. *Blind Insight Blog*, April 2004, http://blindinsight.blogs.com/blindinsight/2005/04/ epiphany.html (accessed 1 Aug. 2005).

21. Our thanks to Christopher Newell for this point, and for a number of other helpful comments on this chapter.

22. Herring et al., "Women and Children Last."

23. For example, http://www.whitestick.co.uk/podcasts.html (accessed 1 Aug. 2005).

24. See http://www.mosenexplosion.com/.

25. See http://www.blindcast.com/podcast/blindspot.html.

26. See http://www.acbradio.org/.

27. See http://jeffbishop.blogspot.com/.

28. See http://www.teleread.org/blind.

29. *Ouch!*, "About Ouch!," http://www.bbc.co.uk/ouch/about/ (accessed 1 Aug. 2005).

30. *Ouch!*, "About the weblog," 2005, http://www.bbc.co.uk/ouch/weblog/bloginfo.shtml (accessed 1 Aug. 2005).

31. *Ragged Edge*, 27 April 2005, http://www.raggededgemagazine.com/life/bloggers0405.html (accessed 5 Aug. 2005).

CHAPTER SIXTEEN

Living in *Cyworld:* Contextualizing Cy-Ties in South Korea

Jaz Hee-jeong Choi

South Korea has been receiving increasing attention from various spheres of international media in recent years, particularly for the nation's exponential growth in the domains of technology and popular culture. As well as leading the broadband world with the highest penetration rate of over 75 percent, the "Korean Wave" is also taking various cultural penetration modes, from technological consumables, such as mobile phones and MP3 players, to more traditional forms of films and television dramas, which make significant contributions to the nation's reported US$650 million cultural export in 2003.[1] However, there is another realm in which South Korea has been strongly expanding: blogging. According to *The Blog Herald*,[2] South Korea is the home ground for three of the world's ten most predominant blog hosts: *Daum Planet Weblog, Yahoo Blog,* and *Cyworld.* Considering that *Yahoo Blog,* the first of its kind launched by the Yahoo! Group, alone currently accommodates over three million bloggers, it does not come as a surprise that a small country whose population exceeds just over 48 million boasts the second-largest number of bloggers in the world, surpassed only by the Unites States. *Cyworld,* whose number of members equates approximately to one quarter of the nation's entire population, clearly leads the blog league within South Korea, while also expanding internationally, as seen in the recent launching of *Cyworld* in China and Japan. This chapter provides contextualization of *Cyworld* in today's South Korean society by introducing *Cyworld* in general, discussing the design of *Cyworld,* and examining some of the major aspects of using, or the user development of, *Cyworld* as an online community.

The Cultural Context of *Cyworld*

While the increasing popularity of blogging has made the word itself a relatively common term worldwide, a different term has been introduced to Koreans to describe a similar, if not identical, concept. In 1999, *Cyworld* launched a service providing an individual online space in which functions similar to those common in blogs were made available to the user. This particular service was named Mini-hompy, a term that was quickly adopted by Korean netizens and used to delineate what would otherwise be commonly referred to as a blog in many countries. Accordingly, it would not be an exaggeration to suggest that the blog culture in Korea was initiated by *Cyworld*. Particularly since 2003, when *Cyworld* became affiliated with SK Communications, itself a subsidiary of SK Telecom Co., Ltd., a predominant wireless services provider with over 52 percent market share in South Korea,[3] *Cyworld* has successfully attracted Korean youth. According to Hyun-oh Yoo, the president of SK Telecommunications, over 90 percent of Korean Internet users in their twenties are members of *Cyworld*.[4]

> **Mini-hompy:** the free, ready-made homepage available to *Cyworld* users, which combines features of a personal blog and an interactive multimedia Website.

Upon acquiring membership, users are given free, unlimited access to their own ready-made online space, called "Mini-hompy." "Hompy" is a common term among Koreans for "homepage" or "Website," therefore the name "Mini-hompy" conveys the notion of an individual's own small online space, rather than a mere journal. This is also visually represented in the fact that the Mini-hompy appears as a smaller pop-up window (920 x 570 pixels) containing eleven default subsections: *Home, Profile, Diary, Mini-room, Jukebox, Photo Album, Gallery, Message Board, Guestbook, Favorites,* and *Administration.* The user can selectively disclose and customize subsections and upload any content that is in accordance with Korean law. Moreover, there is no limit to the total size of the content that each Mini-hompy can contain, although each post can only include files to a maximum of 2MB in size, excluding the actual text content of the post.

South Korea, like many of its neighboring Asian countries, can be classified as a collectivist, interdependent, and high-context culture.[5] Such a culture substantially values the concept of harmony, particularly within one's cohesive and long-lasting in-groups, as the self is not defined solely as an individual entity but rather in relation to the significant others within one's in-groups, such as one's family, friends, neighbors, and co-workers. Consequently, individuals must maintain their heightened awareness of their significant others in order to sustain the duration of and the harmony within one's in-groups, and to thus sustain the individual's own social existence in turn. Furthermore, because communicative norms are high-contextual, such awareness is communicated in an implicit, non-verbalized manner by way of social cues exchange. This means that one is expected to implicitly present social cues to indicate how they are socially situated in a given context, and concurrently evaluate the significant others' reciprocal efforts in the same vein.

Design and Features

Such Korean cultural tendencies within the virtual realm became even more prominent in the late 1990s. The National Internet Development Agency of Korea reports that the annual growth in the number of Internet users varied from 90 percent to 250 percent, while .kr domains flourished with a record growth of nearly 1,000 percent at the peak of this period.[6] Major portals were launched one after another, offering services that showed innate references to collectivist social qualities. The epitome of this trend was *Iloveschool.co.kr*, which became an instant sensation as it provided various means to its members to search for, and communicate with, long-lost friends from school, thereby creating a form of "net nostalgia" for Koreans. *Cyworld* takes a similar approach, as it provides a search function with which the user can search for a particular person either by the combination of name, date of birth and gender, or by email address. However, the uniqueness of *Cyworld* is found in its embracing not only the collectivistic social traditions, but also the increasing individualistic traits of contemporary Korean culture. While *Iloveschool.co.kr* is based largely on the in-group mentality of specific groups (graduates) and limits the online user activities in the "community" space provided for its particular group of users, *Cyworld* is an individual network-based system, giving users the freedom to establish and manage their own personal space in which their online network activities occur.

Several distinct features become evident after a close examination of the design of *Cyworld*. Among those, the features of registration, access methods, *1-chon* relations, and content scraps are particularly notable.

Registration

The registration for *Cyworld* requires input of information similar to that of many Korean online portals. Users are required to provide their names in "real life," in addition to their Korean resident identification numbers, which are then validated online in cooperation with Seoul Credit Rating and Information (SCI) Inc., even before the first phase of the registration process is initiated. Compared with the standard registration procedures of many of the world's leading blog portals, including *Blogger* and *Xanga*, which simply involve the input of a preferred username and a valid email address, the registration process of *Cyworld*, along with many other Korean online portals, appears to be in counterflow to the cyber-utopian vision of "collective intelligence" on the grounds of anonymity as a process of contextual freedom.[7] However, the results of an online poll conducted by *Cyworld* in 2001 reveal that this particular inclination toward offline identity confirmation is not a factor of unidirectional enforcement by the host operator (SK Communications), but a bidirectional agreement between the host and the users: according to the poll results, 36 percent of the participants agreed that their offline identity *should* be disclosed online, while only 12 percent disagreed. Considering that 25 percent of the participants agreed to a partial identity disclosure on certain Websites, including those of government and public services, a total of 61 percent consented to partial or complete identity disclosure, which is over five times the percentage of the opposing opinion. In addition, a considerable number of the participants (26 percent) felt that the issue should be left to the individual to decide, clearly indicating that Koreans, by and large, understand and appreciate the independent and liberating potential of the Internet.[8]

Access Methods

Cyworld currently provides various access methods to the user. Other than access from personal computers, there are two additional methods available for accessing *Cyworld*:

1. *NateOn* (*N@teOn*): a messenger program with similar features to those of MSN Messenger. In addition to conventional messaging functions (chat, email, and SMS) and file sharing, NateOn also features generic functionalities such as the alert service notifying the user of any new posts at their own or their *1-chons*' Mini-hompies, links to commercial gaming and shopping services, as well as mobile phone decoration services (including ring tone and color-ring, a personalized call connection tone heard by the caller).

2. *Mobile Cyworld* (*mCyworld*): reflecting the recent popularity of moblogging, *Cyworld* introduced the *Phone Photo* function in 2003, a function that is essentially identical to moblogging. A designated folder was added as an option to exclusively store and display such images in the *Photo Album* section. Within a few months, a new downloadable program named *mobile Cyworld VM* (*Virtual Machine*) was launched, incorporating the alert function similar to that of *NateOn*, followed subsequently by a new WAP service. With a mobile phone acting as a simplified version of an Internet browser, WAP enables ubiquitous access to *Cyworld*.

1-chon Relation

In Korean society, the term *chon* refers to the level of relatedness within blood-relations. For example, between a legally united couple, there is no *chon* (or *0-chon*), while parents and children have *1-chon* relation. *Chon*-relations are established in a hierarchical manner: a *2-chon* relation is established between grandparents and their grandchildren, while one's aunt/uncle/niece/

> **(Virtual)** *1-chon*:
> the buddy system used within *Cyworld*; *1-chon* relation is established on an invite/accept basis similar to the buddy system of MSN Messenger.

nephew would have *3-chon*, with cousins referred to as *4-chons*. Within *Cyworld*, however, a virtual *1-chon* relation can be created on an "invite and accept" basis. This is similar to the procedure employed in MSN Messenger, in that the user can invite and/or be invited to become a *1-chon*, which the invitee can either accept or reject. If the invitation is accepted, the user and the invitee become "virtual *1-chons*," even though they may not necessarily be *1-chons* in real life. Koreans have traditionally valued cultural homogeneity[9] and have a proclivity toward strong family ties and values.[10] Consequently, use of the term *1-chon* in *Cyworld* conveys a strong, close bond, and even a sense of responsibility and commitment to the user, despite the fact that the *1-chon* relation is established virtually rather than through traditional, non-virtual familial relationships. This is even further enforced by the statistical data provided in the Administration section, analyzing the frequency of interaction between each *1-chon* and the user. In manifestation of such commitment, a new phenomenon has resulted: *1-chon soonhwe*, or "*1-chon* round (tour)" in English, has become a common practice among many *Cyworld* users, who visit *1-chons* on the list one by one to leave a message in the Guestbook section until they have visited every *1-chon*, mainly as a gesture of courtesy.

Content Scrap

One of *Cyworld*'s unique features is the Scrap function, which allows the user to literally "scrap" selected content from a Mini-hompy into their own, excluding comments made about the content in the original site. Technically, this feature shares the characteristics of a deep link, "a publicly accessible HTML 'anchor' tag that points to an off-site web page that is

> **Content Scrapping:**
> a *Cyworld* function that allows users to embed content from another Mini-hompy into their own.

not the home page of the site being pointed to."[11] *Cyworld* takes the middle ground between the positive and negative standpoints of deep-linking by allowing the user to "scrap" content without necessarily obtaining the original author's permission, but simultaneously inserting undeletable hyperlinks to the original site as a form of "reference" or "watermark."

The colloquial expression for Scrap in *Cyworld* is *peogada*, which can be roughly translated as "to draw/ladle/scoop something from a reservoir." This conveys the analogy of drawing water from a well, an abundant source; in this case, making a copy of information that is infinitely duplicable. However, unlike the former analogy in which the act of "taking" does not leave obvious traces regardless of its number of occurrences, in *Cyworld* each content item displays a number that refers to the number of times that the particular content has been Scrapped by other users. A downside, on the other hand, still remains in the fact that there is no option provided for the original authors to track which aspects of their content have been Scrapped, creating potentially significant privacy issues.

Decorating the Mini-hompy

Analogous to various blog portals, *Cyworld* users have an option to purchase decorative items for their Mini-hompies, such as skins (background images), background music, and avatars. It is now a common practice for many Web portals to integrate the use of avatars by appealing to the user as possessing and processing a "visual self," in which they should invest to achieve a better sense of being, similar to the way the self is conceived and constructed offline. On the other hand, the fundamental difference in the use of avatars in *Cyworld* (named "Mini-me"), compared to other portals, is that Mini-me is not merely an image displayed in an empty square area, but one that "resides" in a Mini-room, a semi-3D space for which countless items can be purchased for decorative purposes. For users, this feature creates an improved sense of ownership, hence an increased attachment to their Mini-hompy; for SK Communications, it generates enormous revenue. According to *Hankyoreh 21*, a weekly magazine covering current affairs in Korea, the daily sale of digital acorns, the currency of *Cyworld*, amounts to 150 million won (approximately US$143,000).[12] *Cyworld* provides various methods of payment for purchasing acorns, including credit card, OK Cashbag, and payment as part of a phone bill. OK Cashbag, SK Corporation claims, is Korea's "first and the largest business network forged through business ties with online/offline retailers and mobile service providers."[13] Currently numbered at 20 million, OK Cashbag subscribers can earn points in proportion to their spending at the designated retailers. The accumulated points may then be used as part payment at any of the integrated online and offline business networks.

> **Avatar:**
> a graphical personification of the user in virtual environments such as computer games and online communities.

Figure 16.1: A typical Mini-hompy in *Cyworld*

Another notable feature of *Cyworld* is the user's ability to set the privacy level of the Mini-hompy by disclosing the selected content to the public, *1-chons*, or only to the users themselves. This suggests that the utilization of a Mini-hompy can occur on three different social domains of *global*, *local*, and *individual*. Moreover, fluid access from each of these social domains can only be achieved with the same level of access bandwidth as that of the available information. For example, multimedia content such as an animation can only communicate its intended purpose when the minimum technical requirements are met on both ends of the host and the user before the information itself is communicated. This communication then becomes reciprocal through reflective and reflexive feedback. The design of *Cyworld*, therefore, clearly demonstrates intrinsic considerations toward both socio-cultural and techno-logical parameters of Korean society, which is attributable to the wide appeal of *Cyworld* to the Korean public as a communication system.

Use and Development

The main body of data presented here derives from my previous explorative case study conducted in 2004 on the use of *Cyworld* by Korean youth.[14] One hundred Mini-hompies were randomly selected, to which fifteen analytical questions were applied to obtain quantitative, non-private data. In addition, hybrid email questionnaires were used to acquire more in-depth and individualized information, which resulted in twenty valid responses. Because of the limited space available, this chapter provides insights into some of the more salient and interesting results on users' approaches to the main design elements of *Cyworld*.

Motivation for Registration

According to the questionnaire results, the majority (80 percent) of users attribute their initial acquisition of *Cyworld* membership to "friends-related issues", while some consider "data storage and sharing" as their main purposes. Accordingly, the motivational aspect is discussed from social (friends-related issues) and technological (data storage and sharing) perspectives:

1. *Social motivation*: Korea has the world's highest broadband Internet penetration rate of 73 percent of the total population, combined with the third-largest Internet population in the world.[15] Given that the level of technocapital of a particular society is directly associated with the level of an individual's disposition toward technology, the former (technocapital) and the latter (individual's disposition) inter-influence each other in the course of the collective development of the society.[16] As such, the social and cultural influences on the general Korean public with regard to the use of technology are highly positive, which then results in the public's positive attitude toward new technologies and technological processes, including blogging. Furthermore, as noted previously, long-term human network maintenance is seen as highly important in the collectivistic and interdependent Korean society. Consequently, *Cyworld* immediately appeals to young computer-literate Koreans in a similar manner—not to feel excluded from one's social circles and to keep in contact with them in the virtual realm.

2. *Informational motivation*: Results indicate that a digital camera was the most popular choice of additional technology among the participants, with an ownership rate of 60 percent. In fact, taking digital photos of oneself and uploading such images has become a common routine for many young Koreans in recent years, particularly as a result of the wide availability of digital imaging technologies in cameras, camcorders, and camera phones. One notable example of this phenomenon is the *jjang* (the best) syndrome, which involves online voting by netizens on the uploaded self-photos as a "gateway towards stardom."[17] Different types of *jjangs*, such as *uljjang* (person with the best face) and *momjjang* (person with the best physique) have become "catchphrases in society, entertainment business and other areas" in contemporary Korea.[18] This represents public acknowledgment of techno-social transformations becoming a conventional part of modernization.

Socialization in *Cyworld*

In accordance with the design of *Cyworld*, users' approaches are also profoundly interrelated to their offline social networks. A simple reconfirmation of this aspect can be found in the participants' response to a question on their reason for using *Cyworld*. Eighty-five percent of the participants listed the maintenance and reinforcement of pre-existing social networks as their main motives for *Cyworld* use. New online relations formed in *Cyworld*, however, have rather divergent consequences. Without a single exception, of the 55 percent of questionnaire participants who had formed new social relations through *Cyworld*, all indicated that the relationship was kept strictly online, with one particular participant even labelling the relationship "perfunctory." These findings resonate with Matei and Ball-Rokeach's study on the differential social tie formations of seven ethnic groups in Los Angeles, in which Koreans proved to have more inclination to have associates online, yet simultaneously displayed more caution toward online interaction. The study attributes the in-group cultural orientation of Koreans as the main factor, resulting in their belief that "online relationships outside one's in-group are shallow."[19]

User content development in *Cyworld* shows a similar propensity. A personal blog, in a global sense, may be viewed as a link filter through which personalized information or links to such information are presented, according to the blogger's preferences. Therefore, in order for a blog to enhance its status in the blogosphere, the filtered information it contains must appeal to a large number of audiences to result in a greater number of visitors and potential "secondary authors," compared to countless other blogs available on the Internet. In the case of *Cyworld*, however, the filtered information remains essentially private, as the scale of the potential audience is relatively limited to the members of the user's existing social networks. The questionnaire results provide evidence for this view, with the majority of the participants indicating that their Photo albums and Message boards, the two most frequently used subsections, contain mostly personal data, such as photos of themselves, friends, and family, as well as personal notes and diaries. Therefore, filtering in *Cyworld* remains more social in its nature, rather than informational, as compared to other "mainstream" blogs. Furthermore, precisely because of this social nature of the content in an environment with a dense audience base, the level of Content Scraps remains comparatively high. According to the Mini-hompies analysis, an average of 271 Content Scraps occurred in each Mini-hompy, confirming a high level of information sharing among *Cyworld* users.

Use of decorative items in Mini-hompy is highly conventional (skin: 70 percent; background music: 93 percent; Mini-room decoration: 90 percent), especially through gift-exchange. For an object to be considered a "gift," it should naturally retain a certain value from both the giver's and the receiver's perspectives. Any item purchase transaction in *Cyworld* requires the exchange of currency from Korean Won to the *Cyworld* currency "acorns," as explained earlier. Therefore, there is a definite financial commitment in gift-giving—so, what is the "conceptual value" of such virtual gift-exchange? Among the reasons behind Mini-hompy decoration, recurring motives included "provision of a more pleasant environment for the visitors," "expression of self and change of mood," and "entertainment factors." These responses suggest that there are certain values associated with Mini-hompy decoration, particularly those of entertainment. Mini-hompies are perceived by the users as, first, a "space" of their own; and second, as an "augmented self." Thus, decoration of the user's Mini-hompy is understood as an extended form of self-presentation, combining aspects of interdependence (social cues-provision) and independence (self-expression) within their online and offline social networks.

Attitudes toward *Cyworld*

From the participant responses, it was evident that users find approximately the same amount of advantages and disadvantages in their use of *Cyworld*.

1. *Positive attitudes*: One of the most common beneficial aspects suggested by the participants was the facility to enrich their existing social networks. This aspect of social maintenance is further extended to the redemption of social ties that are potentially at risk of being diminished, lost, or even have previously been lost, particularly because of physical distance. The effortless transcendence of distance through Internet technologies, and its implications in human societies, have been the subject of extensive discussion, particularly in the CMC literature to date. This is also an aspect that has been proven to have an immense appeal to global Internet users, regardless of their ethnic or cultural backgrounds.[20] However, combined with the "user search" function as one of the principal design components, *Cyworld* presents a more convenient way of reconnecting distant and/or long-lost associates. Consequently, it can be argued that this aspect has naturally enhanced *Cyworld*'s appeal because of the collectivistic and interdependent quality of the Korean culture.

   Another significant point to emerge from participant responses was the opportunity for better understanding of the self and significant others through self-expression. Self-expression, or speaking one's mind, is not a traditionally encouraged concept in collectivistic Korean society, as such openness may potentially create individualistic deviance, ultimately resulting in disruption of the group-harmony.

However, the participants not only expressed their appreciation for the opportunities for self-expression that *Cyworld* offers, but one respondent even considered such an experience "therapeutic." These responses clearly signify that there is an evident cultural conversion among Korean youth, a shift toward individualism from the traditional collectivistic conventions. Ito and Okabe's theory of power geometry can be applied in this context,[21] in that *Cyworld* provides an easily accessible digital layer of reality for Korean youth, where interactions can occur regardless of the traditional cultural and social implications. To put it simply, this particular set of results indicates that *Cyworld* is perceived as both a personal space and as an extended self, through which communication occurs under the user's own control.

2. *Negative Attitudes:* Participants expressed concerns with three main types of implied pressures experienced as part of the social adaptation of *Cyworld*. First, the participants commented on being compelled to sign in constantly and utilize *Cyworld* by means of uploading content and monitoring Mini-hompies of their own and their associates. Such utilization of *Cyworld* can be an arduous and time-consuming task, particularly when the user maintains a large number of active *1-chons*. Second, in contrast to their expressed optimism toward *Cyworld*'s basis of offline social ties, participants were also found to experience limitations in terms of the span of their social network. *Cyworld* is used to facilitate the maintenance and enhancement of existing social ties rather than to establish new social connections. For this reason, some participants expressed their concerns about the implications of high social concentration on a limited number of social networks. Third, some participants also commented on the "superficiality" of the self as represented in Mini-hompy. Although Mini-hompy functions as a private space and extended self, through which a new, "free" reality is constructed, the concurrent reality is that the user's awareness of the audience conversely influences the presentation of the user's self in their Mini-hompy. In a way, the self projected in one's Mini-hompy can be viewed not necessarily as a self to *be*, but rather a self to *share*.

Examining interaction within *Cyworld* reveals some emerging patterns analogous to the characteristics of the offline communication culture of Korean society, yet also imbued with unique technological aspects. The extent of *Cyworld* use in Korean youth culture is so exceptional in its frequency and application that it appears to have become one of the standard modes of communication for youth. In support of this, 40 percent of the questionnaire participants indicated that they consider *Cyworld* to be an integral part of contemporary Korean society. The responses of those 40 percent indicated that the main reason behind this view was its extended support for their social endeavors, in addition to the entertainment and therapeutic factors we discussed above.

Overall, then, communication in *Cyworld* takes account of both social and technological contexts of Korean society, with neither aspect having communicative hegemony over the other. *Cyworld* is now, without a doubt, one of the conventional modes of communication in contemporary Korean society, placing on the user a social obligation to stay constantly signed-in; at the same time, it exists not as a completely new or alternative communication system, but as an extension of pre-existing modalities of communication in today's Korean youth culture.

NOTES

1. N. Onishi, "Roll Over, Godzilla: Korea Rules," *New York Times International*, 28 June 2005, http://www.nytimes.com/2005/06/28/international/asia/28wave.html?ex=1277611200 &en=9cc59e6580a62030&ei=5090&partner=rssuserland&emc=rss (accessed 16 July 2005).

2. *The Blog Herald*, "Blog Count for July: 70 Million Blogs," 19 July 2005, http://www.blogherald.com/2005/07/19/blog-count-for-july-70-million-blogs (accessed 10 July 2005).

3. *Hoovers Online*, "SK Telecom Co., Ltd.," 2004, http://www.hoovers.com/sk-telecom/ ~ID__89513~/free-co-factsheet.xhtml (accessed 3 Sep. 2004).

4. Yoo H.O., "Cyworld Storm Heads for Asian Countries," 2005, http://times.hankooki.com /lpage/special/200502/kt2005022320383545250.htm (accessed 11 July 2005).

5. For an extended discussion on these cultural dimensions, see G. Hofstede, "Geert Hofstede Cultural Dimensions," 2003, http://www.geert-hofstede.com/index.shtml (accessed 2 Nov. 2004); H.R. Markus and S. Kitayama, "Culture and the Self: Implications for Cognition, Emotion, and Motivation," *Psychological Review* 98.2 (1991), pp. 224–53; E.T. Hall, *Beyond Culture* (Garden City, NY: Anchor Press, 1976).

6. National Internet Development Agency of Korea, "Number of Internet Users—Yearly Data," 2004, http://isis.nida.or.kr/index.html (accessed 2 June 2005).

7. P. Levy, *Collective Intelligence: Mankind's Emerging World in Cyberspace* (Cambridge, MA: Perseus Books, 1997).

8. For further discussion on democracy and online communities, see H. Rheingold, *The Virtual Community: Homesteading on the Electronic Frontier* (New York: HarperPerennial, 1994).

9. Na E.-Y. and J. Duckitt, "Value Consensus and Diversity Between Generations and Genders," in *The Quality of Life in Korea: Comparative and Dynamic Perspectives*, Sin T.-C.O., C.P. Rutkowski, and Pak C.-M., eds. (Dordrecht: Kluwer Academic Publishers, 2003), p. 418.

10. C.W. Sorenson, *The Value and Meaning of the Korean Family*, 1986, http://www.askasia.org/ teachers/Instructional_Resources/Materials/Readings/Korea/R_korea_13.htm (accessed 28 Sep. 2004).

11. D. Winer, "Deep Linking," 1999, http://davenet.scripting.com/1999/08/09/deepLinking (accessed Sep. 11, 2004).

12. *Hankyoreh 21*, "Cyworld: Rewriting the History of the Internet," 2004, http://h21.hani.co.kr/section-021003000/2004/07/021003000200407140518003.html (accessed 16 July 2004).

13. SK Corporation, "OK Cashbag Service," 2004, http://eng.skcorp.com/intro/in_bizarea/ in_bizarea05.asp (accessed 21 Nov. 2004).

14. J.H. Choi, *Sign In to Cyworld, Move On with Moblog: Blogging the Korean Way?*, unpublished honors thesis (Brisbane: Queensland University of Technology, 2004).

15. Ministry of Information and Communication, *Korea Internet White Paper 2004*, 2004, http://www.mic.go.kr/down.jsp?filePath=/board/eng_data/res_pub_db/&fileName=res_ pub_kwp_2004.pdf (accessed 18 Nov. 2004).

16. V. Rojas, J. Straubhaar, D. Roychowdhury, and O. Okur, "Communities, Cultural Capital, and the Digital Divide," in *Media Access: Social and Psychological Dimensions of New Technol-

ogy Use, E.P. Bucy and J.E. Newhagen, eds. (Mahwah, NJ: Lawrence Erlbaum, 2003), pp. 107–30.

17. Ministry of Information and Communication, *Korea Internet White Paper 2004*, p. 10.

18. Ministry of Information and Communication, *Korea Internet White Paper 2004*, p. 10.

19. S. Matei and S.J. Ball-Rokeach, "Belonging in Geographic, Ethnic, and Internet Spaces," in *The Internet in Everyday Life*, B. Wellman and C.A. Haythornthwaite, eds. (Malden, MA: Blackwell, 2002), p. 419.

20. Matei and Ball-Rokeach, "Belonging," pp. 404–30.

21. M. Ito and D. Okabe, *Intimate Connections: Contextualizing Japanese Youth and Mobile Messaging*, 2004, http://www.itofisher.com/mito/archives/itookabe.texting.pdf (accessed 1 Oct. 2004).

CHAPTER SEVENTEEN

Subcultural Blogging? Online Journals and Group Involvement among U.K. Goths

Paul Hodkinson

Over the past decade, several studies have illustrated the capacity of the Internet to contribute to the development or the reinforcement of music scenes and subcultures. Such studies include Sarah Thornton's identification of the role of Websites as a means of publicizing the details of raves and clubs as part of the U.K. "acid house" phenomenon[1] and Marion Leonard's focus on the role of a network of Web-based "e-zines" (online fanzines) in the development of a global shared identity among Riot Grrrl fans.[2] Other studies have demonstrated the capacity of group discussion forums to facilitate interaction and shared identity among Phish fans,[3] Kate Bush enthusiasts,[4] members of the U.S. alternative country scene,[5] and bedroom electronic music producers.[6] While differing in their specifics, such studies have all demonstrated the capacity for individuals to develop or consolidate their involvement in particular music- and style-related groupings through the use of collective online resources.

My research of the Internet use of U.K. goths during the late 1990s demonstrated findings consistent with such studies. The goth scene has for over two decades been characterized by a strong sense of subcultural community centered upon particular "dark" styles of music and fashion. By the late 1990s, a network of specialist Websites and group discussion fora were playing an extensive role in reinforcing the cohesion and shared identity that characterized the group.[7] In particular, in Usenet groups and email lists, goths found ready-made group spaces that enabled daily access to subcultural information, conversation, and friendships, all of which are liable to intensify their subcultural participation.[8] Importantly, these findings, as with those of other studies, were largely based on the use by music fans of explicitly community-oriented Internet resources. For this reason, I was intrigued to observe that, during the early

2000s, goths collectively migrated the majority of their online communications away from collective spaces, such as discussion groups, in favor of the ostensibly more individual-oriented format of personal journal-style blogs.

Descendants of the phenomenon of the personal homepage, the vast majority of blogs are owned, controlled, and updated by a single individual.[9] An individual emphasis can also be found in the content of many blogs, something particularly apparent in online journal-style blogs (or "lifelogs"[10]), which focus on personal, everyday life and tend to be dominated by "I narratives" rather than discussions of matters of professional or public interest.[11] Yet, at the same time, blogs have become increasingly interactive, to the extent that they are regarded by Herring et al.[12] as a hybrid format, combining elements of the personal homepage with the interactivity of all-to-all communication facilities.[13] Such interactivity has prompted some to ask whether individual blogs might have the potential to form the basis for online communities.[14]

> **Lifelog:**
> a blog whose content is largely focused upon the everyday life of its individual author. Also known as an "online journal."

The blog platform to which U.K. goths transferred the majority of their online communications was *LiveJournal*, a facility associated particularly with "lifelogs" but equally renowned for its extensive interactive features.[15] In addition to comments and inter-blog hyperlinks, *LiveJournal* offers the capacity regularly to read and comment upon the entries of a self-selected list of *LiveJournal* "friends" by way of a single "friends page" that automatically collates new entries as they appear. Users can also participate in subject-based "communities" whose content appears on their "friends page" alongside recent entries to individual journals.[16]

The migration of goths to *LiveJournal* took place cumulatively, a few individuals initially setting up journals and encouraging others to follow suit both through interpersonal conversation and exchanges on goth discussion groups. Once a significant proportion had set up *LiveJournals*, traffic on goth discussion groups receded significantly, prompting even greater numbers to make the switch. In order to assess what role the use of ostensibly individual-centered online journals might play in respect to the participation of individuals in the goth scene, I set up my own *LiveJournal* and used it as a base to read and interact with the journals of goths using the platform. Additionally, I conducted face-to-face interviews with a total of fifteen individuals, all of whom had responded to a request for volunteers posted on my journal. The discussion below illustrates that, although the platform was not overtly community-oriented, the use of *LiveJournal* tended to enhance subcultural participation because it facilitated the development of strong friendships between goths and acted as a valuable means for the transfer of subcultural information, enthusiasm, and commodities.

Subcultural Social Networks

Online journals are often characterized as a form of individual diary and, as such, inward-looking, self-reflective, and private.[17] Indeed, a study by Susan Herring and colleagues found that, compared with other blog genres, online journals tended to involve little if any multi-participant interaction and often did not even utilize reader comment facilities.[18] These findings, from a study that curiously omitted the millions of online journals on the interactive *LiveJournal* platform from its sample, contrasted significantly with the ways in which journals were used by most goths in my study.

As I have argued elsewhere, the use of *LiveJournal* by goths was in some respects significantly more individual-centered than their previous use of collective discussion groups.[19] Most notably, content tended to include a distinctive variety of individual topics rather than being restricted to matters of relevance to the goth community. Meanwhile, although there did exist goth-oriented *LiveJournal* "communities," my respondents used them sparingly at best, preferring the open-ended individual focus of personal journals.[20] Nevertheless, motivations normally associated with diaries, such as keeping a detailed personal record, expressing one's innermost personal feelings, or indulging in private reflection or therapy, tended to be regarded as less important than use of the facility for conversation.

> Sue: Everybody ... likes lots of comments and to start sort of bizarre conversations on their journal—that's why I do it...
> Andy: You want to see how many comments you can get on your post basically isn't it? I find that anyway.
> Sue: ... You're starting a conversation—that's what you're doing. It's the same as any other conversation—it's just that the people you want to converse with are not in the same room...

Although such conversations varied in their depth and subject matter, the desire to generate interaction of some kind at times resulted in the posting of rather trivial or meaningless entries:

> Lorraine: There's [an] impetus to update even if you've nothing to say ... hence posting nonsense... Mostly I post because I feel like yattering.

The importance of interaction and conversation was also underlined by the fact that individuals often spent more time reading and commenting on other people's journals than they did updating their own. One respondent suggested that this made *LiveJournal*'s "friends-list" facility particularly invaluable:

> Jill: People at the click of a button get all the recent posts from all their people they've defined in their lists—they don't have to search around and go to each individual page ... that is the main appeal, because I will read the friends-list more than I will write my own journal.

Friends-List:

an individually selected list of other *LiveJournal* users whose most recent journal entries can be displayed together on a single "friends page." Users can also designate their own journal entries as accessible only to members of their friends-list.

Although it ensured that individuals communicated with a personally hand-picked selection of individuals rather than an all-inclusive "online community," use of the "friends-list" also functioned to enable regular interaction with a relatively stable set of existing friends, rather than random contact with anonymous strangers.[21] Respondents frequently emphasized the value of *LiveJournal* as a means of maintaining regular contact with people they had an existing face-to-face relationship with and to keep up with the various events taking place in one another's lives. Some found *LiveJournal* so useful in this respect that they actively encouraged their goth friends to subscribe to the platform:

> Andy: If there's people I know who don't have *LiveJournal* I start trying to persuade them to get a *LiveJournal*. It's purely because ... it's much easier to keep in touch with people because you automatically ... check it [your "friends page"] everyday ... so you don't have to make an effort [to actively contact people by phone or email].

While this notion of "keeping in touch" often involved long-established friendships, *LiveJournal* was also used as a means to develop relationships with new contacts who had been encountered briefly in face-to-face situations. Compared to more direct forms of personal communication such as email or SMS texting, adding someone to your *LiveJournal* "friends-list" equated to a relatively unintrusive invitation to get to know one another better and often enabled one-off meetings to develop into long-term relationships. Roger explains:

> Roger: There are people I've met briefly in real life and then added them to my "friends-list"—and that's maintained the relationship when had I just met them face to face and not done anything else, the friendship wouldn't have evolved ... . I think that *LiveJournal* friending is less intrusive than an email to someone. It's a good way of sounding people out...

As a result of the number of goths who had *LiveJournal* accounts, the practice of swapping *LiveJournal* user names with people has become an integral and somewhat unique aspect of gigs and club nights associated with the subculture, as in the following example:

> Andy: I think pretty much everyone on my "friends-list" I personally have met in real life ... . You know I met a couple of people I had a laugh with one night at Whitby [goth festival] in November and so it's like "are you on *LiveJournal*," "yeah," "oh excellent right OK," and you write down names on beer mats and stuff...

While *LiveJournal* contact usually took place after meeting at events, occasionally this could work in reverse; individuals who became aware of one another through mutual participation in comments conversations on the journals of mutual friends would subsequently recognize and speak to one another at goth events. The use of user pictures on *LiveJournal* was important to this process of recognition:

> Jill: I saw someone sitting in the corner [of a nightclub] and thought "he looks like the guy that posts to [name] and [name's] *LiveJournals*."... So I went up and poked him ... and he went "yes" and he said "you're a friend of [name's]."... With *LiveJournal* it's different [from email lists] because you've got a little picture that's right by their comments and what they're saying...

As I have outlined elsewhere, the tendency for people to add existing goth friends to their "friends-lists," alongside the identification of other users and swapping of usernames at face-to-face goth events, tended to ensure that use of the platform enhanced the communication of goths with one another rather than diversifying their social networks.[22] The situation described by Jill was a typical one:

> Jill: They [*LiveJournal* friends] usually tend to be the same subculture. Not always, there are exceptions ... occasionally ... you get ... family friends or even brothers or sisters, but ... with my chain of friends they tend to be 95% related to the goth scene... It is cliquey in the sense that goths stick with goths and that—so they're going to be more related to other goths than they are just anybody else that has a *LiveJournal*.

Although goths were not formally bound together within the confines of a shared discussion group, then, the series of "friends-list" links between their individual journals created an identifiable sub-network on the platform.[23] In the following section we will see that, by facilitating regular and relatively insular communication between members of the goth scene, use of *LiveJournal* reinforced their practical and emotional participation in the subculture.

Facilitating Subcultural Participation

Wellman and Gulia have distinguished between superficial "weak ties," which are confined to a narrow shared interest and take place within a single domain, and "strong ties," which involve extensive familiarity and are played out in a variety of domains.[24] Through enabling individual goths to read about and comment upon a variety of aspects of one another's individual, everyday lives, rather than just those aspects di-

Weak Ties:

narrow social relationships restricted to a single shared interest or confined to a particular site of interaction.

Strong Ties:

broad social relationships that involve extensive individual familiarity and are lived out across a variety of sites of interaction.

rectly related to the goth scene, online journals played an important part in the development of strong, intimate relationships between them, which nearly always extended to other forms of interpersonal communication, whether email, online chat, mobile phone, or, most importantly, face-to-face interaction. In turn, the development and/or reinforcement of such strong, multiplex ties between goths served to reinforce participants' general sense of investment in and attachment to the goth scene as a community. As Gareth explains, once locked into such a network of close individual relationships within the subculture, there was greater chance of retaining one's enthusiasm for participating in subcultural activities:

> Gareth: I think it's fairly cohesive—it interconnects people—it connects them together via this thing ... it doesn't really broaden out to people beyond it [the goth scene] ... when people get into goth and then if they get a *LiveJournal* and connect to all their friends, they're more likely to be influenced by those people and there is a tendency more and more to get trapped ... once they're in there's no escape! [laughs]

Most notably, through facilitating such strong individual relationships, *LiveJournal* use encouraged goths to attend more events related to the subculture in order to see one another face-to-face. While this often involved friends located within the same town or city, it was also common for goths to travel to other parts of the United Kingdom in order to go out to events with their *LiveJournal* friends. Meanwhile, as a result of the swapping of *LiveJournal* usernames at goth events, there was a tendency for such trips to result in additional *LiveJournal* contacts—something that in turn would make future expeditions to the same event an even more attractive prospect:

> Roger: If you go to a club one weekend and meet somebody, you'll maintain that relationship [via *LiveJournal*] between visits to the club—so your second visit to the club is going to be a better, stronger experience—because your bonds to everyone else in the club are much stronger than they would have been otherwise.

As well as encouraging general subcultural event attendance by enhancing friendships in this way, the use of *LiveJournal* communication also explicitly encouraged people to go out to pubs, clubs, or gigs that were the subject of journal entries and/or comments. Although personal journals were connected to individually selected lists of friends rather than to every member of a given community, information about goth events tended to spread rapidly from one journal to another throughout the subcultural network. Furthermore, although among my respondents they tended to be used rather sparingly, *LiveJournal* "communities" oriented to the goth scene also often carried such announcements. Below is the text of an entry on the personal online journal of a goth DJ, in which he announced his involvement in a forthcoming event in London:

» Dead & Buried Invasion...

Don Rik and Myself will be guest DJs at the nation's premier Deathrock night - Dead &
Buried - next Friday. So rip your fishnet, big up your hair and come down and show us
your love!
Come down, you know you want to. Jun. 24th @ 12:41 pm

(8 comments | Leave a comment)

Even more significant than general announcements from organizers were
LiveJournal posts from individual participants expressing their enthusiasm
about events, discussing what to wear, arranging to meet people, or merely asking whether their readers were intending to go. More effective than announcements from promoters, this kind of grass-roots collective enthusiasm
not only ensured that readers learned about what was happening, but also created the likelihood that they might become caught up in the enthusiasm and
make the effort to attend. As Stephen put it:

> Stephen: It's a lot easier now to arrange to go out to a club or talk about ... going out
> in Nottingham rather than Birmingham ... it's easier to put up a flag and say "I'm
> interested in this, is anyone else?" And when one person does, somebody else goes
> "oh well if there's a crowd going, then maybe we will as well."

Meanwhile, it was equally common for those who attended events to post enthusiastic reviews, often complete with photographs, during the following few
days, discussing outfits, music, and gossip. The effect of such posts, which often prompted equally enthusiastic comments from readers, was to prolong collective enjoyment of such events and to increase the enthusiasm of readers to
attend them in the future:

> Gareth: You're reading people's *LiveJournals* and they say things like "Well I went
> down Dark Trix [goth night] last night and it was absolutely awesome and I got
> completely hammered and got off with this excellent girl!" [laughs] and you just go
> "I'm going down Dark Trix next time!" so that kind of thing is going to interest
> people and get them more enthused about the goth scene and make them want to go
> out...

LiveJournal use tended also to have the effect of connecting goths to specialist networks of music, media, and commerce. Subcultural information
about Websites, CD releases, fanzines, and specialist retailers would appear at
frequent intervals. In the following exchange, two respondents discuss the
rapid spread of information about a new goth radio show as a result of an initial announcement by one individual:

> Andy: Like [name]'s radio show ... my initial thing was to put it on my *LiveJournal* and
> go "right everybody, this is happening ... try and give [name] a bit of support."
> Veronica: To be fair, from your journal it spread like wildfire.

Meanwhile, *LiveJournal* was regularly used by goths as a means of discussing and exhibiting subcultural fashion and appearance. The ability to include images as part of *LiveJournal* entries and in the form of user icons enabled participants to reveal their appearance to others. While sometimes this was used to present retrospective images of outfits worn to goth events, individuals would also use their journals as a means of consulting with friends as to what to wear and how to do their hair and makeup. It was also common for people to post photographs of recently purchased items of clothing, as explained here:

> Sean: One thing that LJ is better at than mailing lists is that ... I find there's more fashion discussion—people post pictures of fashion and hair and so on ... because it has functionality for photos—people do post photos and you get to see things, rather than going "I got a new pair of leather trousers" ... people will go "I got these new trousers, look, I'll show you a picture of them!"

The ability to exhibit clothing or other elements of one's appearance in this manner served to reinforce the general enthusiasm of both authors and their readers for working on their appearances—something that has long been a central theme of the goth scene. Meanwhile, in a way similar to the exhibition of style at goth events, the consistent elements of goth style on view served to reinforce the consistency of the goth collective style, while the individual variations, the questions asked, and the discussions that took place encouraged those involved to apply a degree of individual innovation and creativity to their own appearance and to draw upon a variety of practical ideas and examples in doing so.

Finally, *LiveJournal* functioned to enhance the direct exchange of music- or fashion-related subcultural commodities. Sometimes those involved in specialist clothing retail would advertise items for sale on their personal journal, but it was equally common for general participants to post lists of personally owned items for sale to the first person to claim them through a comment. As well as helping individuals to accumulate goth clothes and music, this sort of direct exchange of commodities acted as a complement to more general discussions of fashion, music, and events in providing goths with ideas, discussion, and enthusiasm for the styles and practices that made their subculture distinctive.

Conclusion

The communication of goths via personal online journals differed considerably from the "virtual communities" or indeed "virtual music scenes" that have sometimes been identified on group discussion forums and also contrasted somewhat with their own previous use of email lists and Usenet groups.[25] On *LiveJournal*, the majority of interactions took place on a multiplicity of individual spaces rather than a single, all-inclusive community space in which all communication flowed through the center. Yet the intricate connections between the *LiveJournal* use of goths and their existing involvement in the social networks of a multi-domain subculture had led them to utilize the particular mixture of personal and interactive features on offer in a manner that clearly reinforced this subcultural participation. Goths utilized their *LiveJournals* to generate individual-oriented conversation in a way that intensified their networks of strong personal ties with one another and, in so doing, cemented their mutual attachment to the broader off- and online community of which they were a part. Meanwhile, the operation of such friendships, alongside the direct exchange of subcultural information, enthusiasm, and commodities via the platform, served to act as practical means to enhance participation in the core activities associated with the subculture.

NOTES

1. Sarah Thornton, *Club Cultures: Music, Media and Subcultural Capital* (Cambridge: Polity, 1995).

2. Marion Leonard, "Paper Planes: Travelling the New Grrrl Geographies," *Cool Places: Geographies of Youth Cultures*, Tracy Skelton and Gill Valentine, eds. (London: Routledge, 1998).

3. Nessim Watson, "Why We Argue About Virtual Community: A Case Study of the Phish.Net Fan Community," *Virtual Culture: Identity and Communication in Cybersociety*, Steven Jones, ed. (London: Sage, 1997).

4. Laura Vroomen, "Kate Bush: Teen Pop and Older Female Fans," *Music Scenes: Local, Translocal and Virtual*, Andy Bennett and Richard Peterson, eds. (Nashville: Vanderbilt UP, 2004).

5. Steve Lee and Richard Peterson, "Internet-Based Virtual Music Scenes: The Case of P2 in Al.Country Music," *Music Scenes: Local, Translocal and Virtual*, Andy Bennett and Richard Peterson, eds. (Nashville: Vanderbilt UP, 2004).

6. Andrew Whelan, "'Do U Produce?' Subcultural Capital and Amateur Musicianship in Peer-to-Peer Networks," *Cybersounds: Essays on Virtual Music Cultures*, Michael Ayers, ed. (New York: Peter Lang, 2005).

7. Paul Hodkinson, *Goth: Identity, Style and Subculture* (Oxford: Berg, 2002).

8. Paul Hodkinson, "Net.Goth: Internet Communication and (Sub)Cultural Boundaries," *The Post-Subcultures Reader*, David Muggleton and Rupert Weinzierl, eds. (Oxford: Berg, 2003).

9. Susan Herring, Lois Scheidt, Sabrina Bonus, and Elijah Wright, "Bridging the Gap: A Genre Analysis of Weblogs," *Proceedings of the 37th Annual Hawaii International Conference on System Sciences* (2004), http://csdl.computer.org/comp/proceedings/hicss/2004/2056/04/205640101b.pdf (accessed April 2005).

10. Frank Schaap, "Links, Lives, Logs: Presentation in the Dutch Blogosphere," *Into the Blogosphere: Rhetoric, Community and Culture of Weblogs*, Laura Gurak, Smiljana Antonijevic, Laurie Johnson, Clancy Ratliff, and Jessica Reyman, eds., 2004, http://blog.lib.umn.edu/blogosphere/links_lives_logs_pf.html (accessed April 2005).

11. Adam Reed, "'My Blog Is Me': Texts and Persons in UK Online Journal Culture (an Anthropology)," *Ethnos* 70.2 (June 2005).

12. Herring *et al.*, "Bridging the Gap."

13. Graham Lampa, "Imagining the Blogosphere: An Introduction to the Imagined Community or Instant Publishing," *Into the Blogosphere*, Gurak *et al.*,2004, http://blog.lib.umn.edu/blogosphere/blogs_as_virtual_pf.html (accessed May 2005).

14. Anita Blanchard, "Blogs as Virtual Communities: Identifying a Sense of Community in the Julie/Julia Project," *Into the Blogosphere*, Gurak *et al.*, 2004, http://blog.lib.umn.edu/blogosphere/blogs_as_virtual_pf.html (accessed April 2005); Carolyn Wei, "Formation of Norms in a Blog Community," *Into the Blogosphere*, Gurak *et al.*, 2004, http://blog.lib.umn.edu/blogosphere/formation_of_norms_pf.html (accessed April 2005).

15. Clay Shirky, "A Group Is Its Own Worst Enemy," *Clay Shirky's Writings about the Internet* (2003), http://www.shirky.com/writings/group_enemy.html (accessed Oct. 2003).

16. Paul Hodkinson, "Interactive Online Journals and Individualisation," *New Media and Society* (forthcoming in 2006).

17. Sandeep Krishnamurthy, "The Multidimensionality of Blog Conversations: The Virtual Enactment of September," paper presented at the Association of Internet Researchers Conference 3.0, Maastricht, The Netherlands, 2002.

18. Herring *et al.*, "Bridging the Gap."

19. Hodkinson, "Interactive Online Journals and Individualisation."

20. Hodkinson, "Interactive Online Journals and Individualisation."

21. Hodkinson, "Interactive Online Journals and Individualisation."

22. Hodkinson, "Interactive Online Journals and Individualisation."

23. Hodkinson, "Interactive Online Journals and Individualisation."

24. Barry Wellman and Milena Gulia, "Virtual Communities as Communities: Net Surfers Don't Ride Alone," *Communities in Cyberspace*, Marc Smith and Peter Kollock, eds. (London: Routledge, 1999).

25. Watson, "Why We Argue about Virtual Community"; Lee and Peterson, "Internet-Based Virtual Music Scenes"; Hodkinson, "Net.Goth."

CHAPTER EIGHTEEN

Fictional Blogs

Angela Thomas

F ictional blogs have been lauded by journalists as a "hot new literary trend that has revolutionised publishing."[1] A fictional blog can be defined as any form of narrative that is written and published through a blog, *LiveJournal*, or other similar online Web journal. To date there have been a number of blogs that have been published in the mainstream print media, some of them nonfiction (such as *Rebecca's Pocket*[2]), some fiction (John Scalzi's *Agent to the Stars*[3]), and others whose real/fiction status is ambiguous and contested, such as the *Belle de Jour*[4] (a blog about a London call-girl). Although the claim that fictional blogs are revolutionizing publishing might be overstating the current situation, fictional

> **Fictional Blog:**
> any form of narrative that is written and published through a blog, *LiveJournal*, or other similar online Web journal.

blogs are nevertheless an emerging phenomenon with much potential, particularly for younger and emergent writers. Blog fiction is a way for budding authors to experiment with their style, find their own narrative voice, and workshop their ideas with the possibilities of audience feedback. According to Faleiro, for authors of blog fiction, "the blogosphere thus acts as a cocoon; a space where writers feel appreciated and encouraged, and can identify and define their skills."[5]

The various forms of blog fiction are delineated in figure 18.1. This sets up a clear distinction between (a) blogs that are used simply as a means for publishing a narrative in serial format (one post at a time), and (b) blogs that are used more as a device for writing, where the author(s) actually incorporate the features afforded to them in the blogging software and the online environment. Blogs that are used as a writing device may contain the entire story world within the blog fiction. Alternatively, they may only contain a part of the story world. Finally, fictional blogs have become a popular way to market various commercial products, such as shoes (*Manolo's Shoe Blog*[6]) and even radio shows (*Morris Telford: The Blog*[7]).

Typology of Blog Fiction			
BLOG USED AS A PUBLISHING TOOL • using the blog or journal as a means of publishing writing • these may be original works (*Entia*[8]) • some people have blogged novels that are written in diary form (such as Mary Shelley's *Frankenstein*) as a form of online reading group (*Dracula Blogged*[9])	**BLOG USED AS A WRITING TOOL** • using the blog as a writing device, taking into account and manipulating various or all of the features of a blog: hyperlinks, images, the comment feature, and so on		
	Contained Story • the story world is contained within the blog itself (*The Glass House*[10])	**Partial Story** • the story world is only partially represented through the blog, and is attached to a forum or other community, often a fan fiction community	
		Interactive Role-Playing • this interactive blogging resembles role-playing games, but there are some sets of blogs that link to the role-playing forum as part of the storytelling (*Providence*[11])	**Character Diary** • blogs or journals that are told from a character's perspective, used either to create back story or to integrate within fiction or nonfiction from outside the blogosphere.
			Fictional Source (*Buffy the Slayer*[12]) · **Nonfiction Source** (*Bloggus Caesari*[13])
	FICTIONAL BLOGS USED FOR COMMERCIAL PURPOSES • *Morris Telford: The Blog*[14] (an amusing adventure diary, by the BBC) • *Barbie's Blog*[15] (the life and times of Barbie, by Mattel) • *Manolo's Shoe Blog*[16] (shoe recommendations by a fictional Manolo)		

Figure 18.1: Typology of Blog Fiction

Blog Used as a Publishing Tool

In some forms of blog fiction, an author or authors publish a chapter or part of a story they are writing on a regular basis on a blog. For these authors, the blog is simply a *publication* device and could in effect be exchanged for any other form. Serialized fiction is not something new and exists in a range of media forms, such as newspapers, magazines, and television. One example of this is *Entia*,[17] a science fiction novel in the process of being written by Jay and David Steele. The authors publish one chapter at a time and elicit reader feedback through the commenting system of the blog.

Works of fiction (either original or classic) published on blogs for the purpose of soliciting audience feedback are being met enthusiastically by small groups of readers. It does seem, however, that although the medium of publication is new, there is nothing particularly unique in the writing or narrative form. Instead, what is unique is the quick and easy blog commenting mechanism for relatively instant feedback and critique.

> **Blog Fiction Comments:** the quick and easy commenting feature of blogs allows relatively instant feedback, encouragement and critique from readers; this is a unique feature of blog fiction.

There are also existing works, such as Bram Stoker's *Dracula*,[18] that have been blogged (*Dracula Blogged*[19]) for the purpose of creating an online reading group. In this instance, the blogging tools serve both as a means for serial publication and for readers to use the blogging comment facility to discuss the finer points of the novel.

Blog Used as a Writing Device

Blog fiction as a genre might best be defined as that fiction which is produced where an author or authors have used a blog as a *writing* device, using all of the features afforded by the blogging or journaling software, such as hyperlinks, graphics, and the commenting system. Such authors are experimenting with and manipulating the software to exploit it for their writing purposes, and in this way *are* creating a genre that has both continuities and disruptions with other narrative genres. Their innovative play with the medium is creating a narrative, which at its best is multimodal, hypertextual, episodic, serialized, and interactive. Commenting about the interactivity in the blog novel of *Diego Dovel*, Jim McLellan reported that

> Blogging's sense of immediacy was key. He didn't plan a story in advance, but improvised each day. Though readers don't "direct" the story, the response from them every day probably did have an effect. "When you're writing, there is a kind of idealised reader in your mind. Here, the idealised reader becomes very real. It's all these people sending email and commenting on their own blogs."[20]

The impact of interactivity on writing is crucial to the unique experience of creating blog fiction.

Full Story World Self-Contained in the Blog

In the typography, a distinction is made between blog fiction in which the entirety of the story world is self-contained in the blog, and where the story world is only partially represented. *The Glass House*[21] is an excellent example of a fully contained blog fiction that manipulates the blogging software for all its affordances, thereby creating a narrative that is characterized by the use of multimodal, hypertextual, episodic, serialized, and interactive features. In *The Glass House*, the reader gets to see inside the life and mind of James. And James is invisible. In the first entry, the reader is provided with an overview of the story:

> I'm James. If you aren't already in my circle of unusuals, here's what's important:
>
> **I'm invisible.** I mean this literally. It's not a metaphor; it's not a bitch about my social life. I mean that light does not reflect off of my body, nor things under my influence.
>
> **Most of my friends are stranger than me.** That's how it goes. I know two telepaths. My best friend's a gay telekinetic. My housemate's unstuck in time. My girlfriend (I think it's safe to call her that) warps spacetime. And those are just some of the people whose faces I know. I've got weirder acquaintances online, though some of them of course I just don't believe.
>
> **We, ahem, fight crime.** Sometimes. To be fair, sometimes we commit crime, though none of the really bad ones so far. And sometimes we use our power just to get the good seats at the movies.
>
> There are consequences, of course. Inquiries. Enemies. Temptations of power. I'll talk about those. But it's the day-to-day stuff you'll hear about most. Few things are really ordinary when you can't order a burger at McDonald's without spooking people. This is not some superpower I gained from a comic book genesis story. Nothing radioactive bit me. I was born this way. I can't turn it off.[22]

The writer, James, uses standard literary devices such as flashbacks and allusions to explain and suggest the state of his current life (such as the hilarious line: "invisible people should not work with hand tools"). James also directly addresses the reader at points in the narrative ("imagine, unknown reader, that you were possessed with the miraculous power to make people happy"), making meta-fictive comments as well as instructing the reader to be patient with his storytelling. He also occasionally includes out-of-character comments in parentheses to explain or apologize for being disrupted in the middle of his writing. The characters are fleshed out and real, they yell at each other over mundane things, yet they truly care for each other despite, or maybe because of, each other's "unusual" qualities. Readers quickly fall in love with James during his successive hilarious entries, from scaring a young boy on a bus, to playing mind tricks on a group of fourteen-year-old girls as they attempt to perform a Wiccan ritual during a sleepover party.

However, what also makes the writing distinctive as well as incredibly amusing is that he uses the commenting system of the blog to leave fictional comments by his friends, and this not only exploits the potential of blogging for this purpose to its fullest, but it cleverly adds another layer to the narrative. Each of the fictional commenters adopts a particular style as well, and the repetition of this style constructs an image of their persona. And when real people leave comments or questions about the story or seeming incongruences in the story, James weaves the answer or explanation into a later post.

Sometimes in his posts, James talks directly to one of his friends ("to Dana: No, I'm not going to be on IM"), and as the narrative progresses, the "blog-ness" of it begins to emerge more, with the use of hyperlinks to songs, the *Wikipedia*, poetry, books he is reading, news items, and even to the posts of other bloggers. James becomes more playful with the genre as he progresses, too, including email excerpts, movie reviews,

> **Benchmarks for Blog Fiction:** narratives that exploit all of the features of blogs to construct the world of the narrative are defining benchmarks for blog fiction. Such features include the use of images, audio files, hypertext links, and the audience commenting system.

his own poetry, chat transcripts, and images. We are also privy to the occasional guest posts by his (fictional) girlfriend Callie. The exploitation and manipulation of the blog features in this blog fiction is playful and innovative, and defines a standard for blog fiction.

Partial Story World Contained in the Blog

A further type of blog fiction identified in the typology is one where only part of the world of the story is represented in the blog. In some instances, groups of writers define a story world together and then individually write concurrent blogs from the perspective of each of the characters in that world. This idea is articulated by fiction blogger Jack Chan, who explains:

> you can have multiple characters, each with their own blogs, creating this intertwined story with plenty of mystery, cliffhangers, and plot twists. With every new post, the story progresses, and even the most mundane entries develop the characters. The authors become actors, reading what the others write and then adapting the story accordingly. Romance. Suspense. Comedy. You name it. It's part creative writing, part blogging, and part role-playing.[23]

This concept is popular in the fan fiction writing community, and there are a number of role-playing/fan fiction-style *LiveJournals*, where writers (often younger writers) blog as fictional characters. From the fan fiction community of *Buffy the Vampire Slayer*, for example, various young writers are keeping journals in the role as the characters Buffy, Giles, Xander, Willow, Dawn, and Clem (see associated URLs below[24]). Similarly, there are groups of journals for the characters in *Charmed*[25] (e.g., Cole Turner) and for the television show *Providence*,[26] which also has an associated Website and role-playing forum. Additionally, fans of certain historical figures have used them as a basis for creating a fictional blog based on their lives. *Bloggus Caesari*,[27] for example, was a blog written to record the life of Julius Caesar in diary format.

Fan Fiction Blogs

A previous study conducted by the author[28] examined the fan fiction practices of a group of adolescent girls. The fan fiction that the girls co-created in this study was a form of crossover fan fiction—they combined the worlds of *Star Wars* and Middle Earth as well as alternative universes to develop their plots. The girls role-played both synchronously (through Yahoo! Instant Messenger) and asynchronously (through their role-playing forum, *Yoda Clones*[29]) to create the basis for their fan fiction. One of the girls would then rewrite those transcripts into a single fan fiction text to post it at *Fanfiction.com* in order to get feedback and a wider readership. In 2004, however, several girls decided to start a *LiveJournal* and a blog for their central character for the purposes of exploring and developing their character and enhancing their fan fiction writing through better characterization.

> **Fan Fiction:**
>
> the practice of innovating on a much-loved story to develop alternative versions of that story. Fan fiction has flourished with new online fan communities, and is popular with adolescent writers.

In an article about the benefits of fictional blogging for teens, Silvester of the e-zine *About* suggested that

> this is a great way to tell stories, and to get your stories out there to a wide readership. It also seems to me to be a great way to explore characters. Even if you can't make a publishable novel out of a Weblog, writing one from the point of view of a character would really let you get into that character's head. Then it would be even easier to use the character in other forms of fiction.[30]

In addition to getting inside a character's head and creating a back story for fan fiction writing, these particular online journals are also a means of exploring and constructing the self, and the girls mentioned in the above study were authoring versions of themselves as they write in role. It was found that the narrative and fiction served as a safe distancing mechanism to explore feelings and experiences of adolescence that were either difficult or unexplored through their real selves.

A common practice by fan fiction writers is to insert versions of themselves into their characters, which is known variously as "fusing identities," "hybrid identities," or creating a "Mary Sue" character.[31] Tiana and Jandalf, two of the girls in the study, both openly stated that their characters were very much adaptations of their own identities, made all the stronger through both the role-playing, which relies upon a considerable degree of instinctuality, and the *LiveJournals*, which allow a more introspective reflection on ways in which their characters might be facing issues and angst-ridden

> **Fictional Blogs as Identity Performance:** adolescent writers of fictional blogs in this study used the central character of their blogs to explore aspects of their own identities: their adolescent excitement and angst, difficult everyday experiences, the very real pains of growing up, and possible versions of their future selves.

insecurities similar to what they are facing in their real lives.

The fusion of identity is clear in Tiana's fictional blog. Tiana often discussed the ways in which her character dealt with adolescent issues such as growing up and peer pressure. One particular poem she wrote in her journal exemplified this. A close linguistic analysis of this poem reveals how Tiana's language choices vividly construct the frustrations and angst experienced by her character and, in turn, her real self. Tiana explained that the nature of the poem had to do with what her character was going through, in fact describing it as an "angst-drama-type" poem.

> **» Calling**
>
> I'm stuck within a tangled web of mist and betrayal,
> Looking in a mirror of me, I wonder of this portayal.
> Who is this shadow that I see,
> This isn't how this has to be...
>
> Standing here, I wonder how to take this,
> No longer can I see myself in happiness or bliss.
> Once upon a time I loved,
> What's happened to my beloved?
>
> There's nothing left but answerless questions,
> What's happening, do you have any suggestions?
> I'm losing myself,
> *You're* losing yourself...
>
> Don't make me lose what I've only just gained,
> Can't you see the expressions, all so pained?

Once I drew away from all,
At least I returned at your call.

Why can't you hear me when I cry,
Why can't you hear my heart's sigh?
Don't leave again,
Must I remain?

Alone in the darkness I walk once more,
I'm searching for a way to open the door...
The Light is calling...
Again I'm falling...

Even the darkest shadow can be reborn,
But now I'm stuck between the two—torn.
Is this then the end?
Don't leave me, my... friend.

Tiana explained that the poem was about her character struggling again with her dark side and calling out to her Jedi Master, Jandalf. Both Tiana and Jandalf used their *LiveJournals* and blogs to recount early character memories. By entering character memories into their Web journals, the girls are forming that fictional identity while at the same time situating that character into a particular family, time, place, and community. Furthermore, through inserting their own real memories into their character's journal entry, they are fusing together the two identities, real and fictional, creating a hybrid person who is at the same time them and yet not them.

The fictional characters who are central to the girls' blogs are also a means for them to fashion new and emerging identities for themselves as they develop into adulthood. The characters allow the girls a freedom and power to "author an identity"[32] that plays out their fantasies and desires: of their physical bodies, their hopes and dreams for the future, and their ideas of romance. Their characters are a rehearsal of who they want to become, and in role-playing that ideal self, they can grow closer to becoming that ideal. It is the imaginative possibilities of their fictional characters that empower the girls' ability to imagine these same possibilities for their real selves.

Tiana reflected on the dialectic nature of her real self and her characters, stating:

> I model bits of myself into my characters by just letting go, per se. It's really the other way around: I infuse the characters into myself, more. You let the characters become a part of you, let yourself be able to think like they would, and it works the other way

around. You can't have a character that doesn't have some of your personality without losing yourself.

Tiana's statement provides a lucid insight into the blurring of the fictional and real spaces in which she exists, both as herself and as her characters. She talks of always being "plugged in" to her characters, and as a consequence her real self is blended with the characteristics of the fictional Tiana. Jandalf, too, talked about the ways in which her character influenced her own sense of self while impacting her real identity, so much so that she claimed, "I've found that since I've been using her as one of my main characters, I have been ... well ... rubbing off on myself, in a way. I'm more outgoing than I used to be, and Jandalf's creation and use does figure in that." What both girls have done in fact is to write themselves into a new identity, empowering their realities through their fiction.

Fictional Blogs Used for Commercial Purposes

The final form of fictional blog in the typology is that of the commercially produced blog to support a franchise (such as Mattel's fictional *Barbie's Blog*), a television show (such as the BBC's *Morris Telford: The Blog*), or a more general product (such as *Manolo's Shoe Blog*). The recognition of blogs as a marketing tool is also recognition of the significance of blogs and the power of blog fiction to reach media-savvy audiences. The curiosity sparked by the clever wit and humor found in *Manolo's Shoe Blog* has led to the person behind it becoming something of an Internet legend. While the character is fictional, the witticisms are heartfelt, and readers are never quite certain where the fiction ends and the reality begins. Furthermore, readers are entertained by both the fiction and the reality without ever being overly forced to engage in the commercial subplot of the blog. Intense and intelligent debates about fashion are played out in the comments of the blog, combining topics such as fashion, identity, and Foucault. Yet the fictional threads ("the" Manolo's life) are woven very tightly into the blog to keep it entertaining. This is a very innovative approach that blends genres (the social purposes) within the blog to create something unique, amusing, and commercial all at the same time.

Conclusion

Fictional blogs have provided innovative writers with new ways of exploring narrative form, including the use of text, images, audio, and hypertext. For me as an English educator, it is particularly exciting to consider the opportunities for fictional blogs such as character diaries to give emergent writers the temporal space to develop their own narrative voice, the interactive space to give and receive feedback, and the identity space to explore their emerging lives as young people. As Gauntlett argues, "to interpret the choices we have made, individuals construct a narrative of the self, which gives some order to our complex lives."[33] He also claims that in forging our identities, we construct a narrative in which we play a heroic role. The fictional characters played by Tiana and Jandalf allowed the girls to exaggerate that heroic role and to play with the fantasies of who they want to become. By finding and developing a narrative voice, they let us glimpse their real thoughts and feelings. For Giddens,[34] the narrative self of women was traditionally caught up in romantic storylines. However, Tiana and Jandalf are writing their own science fiction narratives, in which they are Jedi knights, fighting with light sabres, and are active agents of their own futures. The process of discovering their character is also a journal of self-discovery, for understanding their pasts and for forging new identities for their futures.

Finally, fictional blogs have also given rise to many other variations—from fans who want to be involved in both the interpreting of and production of blogs related to film, television, radio, and printed texts, to the promotion of commercial products. One major shift in thinking that this brings about is that we can no longer think of readers as passive recipients of media, but as active respondents and producers of their own versions of fictional texts. As also noted in the introduction to this collection, we can no longer think of consumers of commercial products as passive; instead they are actively debating the merits of products, even within fora that are blending the real products within a fictional context, such as *Manolo's Shoe Blog*. As Wright argued, "as more and more people start blogging, the lines will inevitably blur between author and reader, and between fact and fiction."[35] Fictional blogs in their many variations have much potential and may indeed become, in Faleiro's terms, revolutionary, not just in publication terms, but for all their interactive and innovative narrative possibilities.

NOTES

1. Sonia Faleiro, "There Is Someone Out There," *The Sunday Express*, 16 May 2004, http://www.indianexpress.com/full_story.php?content_id=46845 (accessed 17 Mar. 2005).

2. Rebecca Blood, *Rebecca's Pocket*, http://www.rebeccablood.net/ (accessed 17 Mar. 2005).

3. John Scalzi, *Agent to the Stars*, http://www.scalzi.com/agent/ (accessed 17 Mar. 2005).

4. *Belle de Jour*, http://belledejour-uk.blogspot.com/ (accessed 17 Mar. 2005).

5. Faleiro, "There Is Someone Out There."

6. Manolo, *Manolo's Shoe Blog*, http://shoeblogs.com/ (accessed 17 Mar. 2005).

7. BBC, *Morris Telford: The Blog*, http://www.bbc.co.uk/shropshire/features/blog/archive.shtml (accessed 17 Mar. 2005).

8. Jay and David Steele, *Entia* (2004-2005), http://entia.typepad.com/ (accessed 17 Mar. 2005).

9. *Dracula Blogged*, http://infocult.typepad.com/dracula/ (accessed 17 Mar. 2005).

10. *Glass House*, http://www.invisiblejames.com/ (accessed 17 Mar. 2005).

11. *Providence*, http://www.geocities.com/providencemods/providence/index.html (accessed 17 Mar. 2005).

12. *Buffy the Slayer*, http://www.livejournal.com/users/xbuffysummersx/ (accessed 17 Mar. 2005).

13. *Bloggus Caesari*, http://www.sankey.ca/caesar/index.html (accessed 17 Mar. 2005).

14. BBC, *Morris Telford: The Blog*.

15. Mattel, *Barbie's Blog*, http://myscene.everythinggirl.com/friends/barbie/barbie_index.aspx (accessed 17 Mar. 2005).

16. Manolo, *Manolo's Shoe Blog*.

17. Jay and David Steele, *Entia*.

18. Bram Stoker, *Dracula* (London: Archibald Constable Company, 1897).

19. *Dracula Blogged*.

20. Jim McClellan, "How to Create a Blog-Buster," *Guardian Unlimited*, 10 Apr. 2004, http://www.guardian.co.uk/online/weblogs/story/0,14024,1187641,00.html (accessed 17 Mar. 2005).

21. *The Glass House*.

22. *The Glass House*.

23. Jack Chan, "Blog Fiction," *Blogfic Forum*, http://www.blogfic.com/forum/ (accessed 17 Mar. 2005).

24. *Buffy*-related blogs (all accessed 17 Mar. 2005):

    Buffy   http://www.livejournal.com/users/xbuffysummersx/
    Giles   http://www.livejournal.com/users/xrupertgilesx/
    Xander   http://www.livejournal.com/users/xander__harris/
    Willow   http://www.livejournal.com/users/red_witch/
    Dawn   http://www.livejournal.com/users/xdawnsummersx/
    Clem   http://www.livejournal.com/users/clemtastic/

25. *Cole Turner*, http://www.livejournal.com/users/cole_turner/ (accessed 17 Mar. 2005).

26. *Providence.*

27. *Bloggus Caesari.*

28. Angela Thomas, "Blurring and Breaking through the Boundaries of Narrative, Literacy and Identity in Adolescent FanFiction," paper presented at National Reading Conference, Miami, December 2005.

29. *Yoda Clones,* http://yodaclones.proboards29.com/index.cgi (accessed 17 Mar. 2005).

30. Niko Silvester, "Fictional Blogging," *About,* 10 Mar. 2004, http://teenwriting.about.com/b/a/071172.htm (accessed 17 Mar. 2005).

31. Rebecca Black, "Access and Affiliation: The New Literacy Practices of English Language Learners in an Online Animé-Based Fanfiction Community," paper presented at 2004 National Conference of Teachers of English Assembly for Research in Berkeley, CA, p.1.

32. Michel Bakhtin, *The Dialogic Imagination* (Austin: U of Texas P, 1998).

33. David Gauntlett, *Media, Gender and Identity* (London: Routledge, 2002), p. 113.

34. Anthony Giddens, *Modernity and Self-Identity: Self and Society in the Late Modern Age* (Cambridge: Polity, 1991).

35. Tim Wright, "Blog Fiction," *trAce,* 16 Jan. 2004, http://trace.ntu.ac.uk/Process/index.cfm?article=91 (accessed 17 Mar. 2005).

SECTION THREE
OUTLOOK

A Vision for Genuine Rich Media Blogging

Adrian Miles

Blogs are a rich, diverse, and quintessentially disparate medium expressing the Internet as a network of noise, connection, communication, and difference. Lately these qualities have become more evident with the appearance of traditional time-based media in blogs, principally audio and video, and their more recent progeny, pod- and video-"casting." The incorporation of audiovisual media within blogs has seen the development of substantial new blogging genres and has the potential to generate new genres of audiovisual content and associated technologies. However, the key problem confronting the successful incorporation of audio and video into blogging practice revolves around how those qualities that make a blog a blog can become part of time-based media, as opposed to an appropriation of blogs as mere distribution or publishing "engines" for audio and video files.

> **Podcasting:**
> a way to distribute audio and video content via the Internet; users subscribe to "feeds" pointing to media files which can be downloaded automatically and transferred to portable iPod or MP3 players.

There are, as this volume indicates, many ways in which blogs can be defined and theorized. The contribution I wish to make to this discussion is to identify blogs with those formal features of blog Content Management Systems (CMSs) that can be seen as a material response to the "affordances" of networked writing. Affordance is a term popularized by the industrial designer Donald Norman and refers to a user's perception of what can be done with an object.[1] In the case of blogs, the generic (and hegemonic) form in which blog software has developed "affords" such forms as the writing of individual posts that have a heading, date, and time stamp, the automatic attribution of authorship, the optional provision of comments, the archiving of posts based on category and date, and the automatic provision of a permanent URL for individual entries.

Additionally, blogs have also inherited much of the affordances of hypertext, evidenced in the manner in which their basic unit of construction is the post, which is essentially a small chunked hypertextual node. This node is able to be read and understood on its own (generally it is not necessary to read an entire blog to understand a single entry) and by virtue of its permalink can be interwoven hypertextually with other nodes. Whether entries are read in the context of an original blog post or whether they are viewed by way of other blogs hardly matters.

Another series of affordances are realized as a consequence of the networked nature of blogging (though, of course, the hypertextual and networked nature of blogging means that these two key attributes are deeply intertwingled), and this is evident in how blogs generically contain blogrolls, trackback, RSS, permalinks, and also the increasingly common provision of links to third-party blogosphere, folksonomy, or social software sites such as *Technorati*, *Blogstreet*, *Flickr*, *Del.icio.us*, and *Blogshares*.

> **Content Management System (CMS):**
>
> the Website backend softwares that handle the publication of blogs and other forms of collaborative online publishing, reducing or even eliminating their users' need to be familiar with HTML mark-up.

In general, these generic attributes can be understood as a feature of blogging as a networked hypertextual writing activity, occurring directly as a result of the material technological affordances of specific CMSs. These tools make certain sorts of writing possible, particularly a writing that is beyond or outside of writing narrowly conceived as *my* words on *my* screen, and they form the foundation of blogging as a medium.[2] In addition, blogging also expresses many of those qualities that were originally attributed to hypertext more generally. For example, the blogosphere is multivocal, multilinear, and has moved past print to produce complex intertwingled docuverses of interconnecting fragments.[3]

Toward Audio and Video Blogging

Many of these qualities are also utilized in audio and video blogging; however, it is also apparent that in present practice much of what can be characterized as the basic affordances of blogs are lost, or ignored, in audio and video blogging. To illustrate this, we can perhaps use the recent and explosive development of audio blogging. The ability to embed audio in a Web page (as opposed to making an audio file available for download and playing in a separate player) has been available since 1996, when Apple first released a browser plugin that supported its QuickTime format. However, it was the development of podcasting clients in 2004 that seeded the rapid and exponential rise of audio-enabled blogs. These clients enabled users to subscribe to RSS feeds that contained audio "enclosures" in exactly the same way that RSS aggregators facilitated the rise of RSS as a major distribution form—indeed, pod- and videocast clients are essentially RSS aggregators that support media enclosures. In the case of podcasting, as the term indicates, the best clients automatically synchronize these audio files into Apple's iTunes library and place them onto users' iPods and other portable MP3 players for later listening. With the rise of the Video iPod, exactly the same can now be done for video files.

> **Media Enclosures:**
>
> pointers within an RSS feed that locate media objects, such as audio and video files, and download these in the background.

RSS feeds, which have driven the success of pod- and videocasting, are generally automatically produced by blog CMSs, and where they are not, several third-party services are available to produce appropriate RSS feeds. It is these feeds that users subscribe to, and in this manner audio and video files are distributed to clients. This aspect of pod- and videocasting clearly takes advantage of blogs as distributed personal publication and distribution technologies, and has successfully appropriated the lightweight protocol of RSS to provide the infrastructure for the development of an alternative distribution regime. This is impressive and has led to a rise in audiovisual "prosumer" commentary on a variety of topics—particularly in podcasting, where the best content is (as with blogging more generally) on a par with any commentary heard on public radio. At the same time, it is also true that the worst audio and video blogging content is, frankly, deplorable; this is, after all, the consequence of any distributed and accessible networked technology that allows individuals to become media producers and distributors.

This audio and video blogging content is as diverse in style, content, presentation, and technical excellence as is the prose delivered in text-based blogging. It comprises pieces produced to professional or near-broadcast standards, through to what can generously be described as naïve media works. However, while this diversity of content and style is a feature that audio and video blogging shares with traditional blogging, there are many other important facets of text-based blogging that this rich media blogging does not yet achieve. Most of the qualities that make a blog *a blog* have been translated into the content, but the specific networked and hypertextual affordances of textual blogging have not yet significantly penetrated the field of audio and video blogging. This is, of course, why the suffix *casting* has been so successful and intuitive a descriptor for those undertaking these activities, as this rich media blogging is a practice that looks more to old media models than to the affordances and possibilities already realized and provided in what is now the canonical model of text blogging.

Current Limitations of Audio and Video Blogging

Let us critique in more detail aspects of existing practice, before proceeding to a discussion of other possibilities and futures. As a video blogger, I certainly don't believe that the revolution has yet happened, in spite of the runaway success of podcasting and the rapidly pursuing videocasting. To date, the major achievements of both these rich media forms of blogging are best celebrated and understood in the light of existing media institutions and traditional mass media. As with traditional text-based blogging, it was not that long ago that to have a publication with an international audience would require a very substantial capital outlay. Even if self-publishing, the costs associated with writing, editing, printing, distribution, advertising, and sales are potentially enormous, and so have always acted as a barrier to entry. This is, of course, one of the reasons why mass media developed in capitalist economies— audiences must be maximized to generate a return on such capital outlay. The emergence of the World Wide Web dramatically affected publishing and production costs; suddenly anyone with a Website had the capacity to write and distribute their work for negligible costs internationally. Blogs have taken this publishing ease a step further, allowing any consumer of Internet content to become a site publisher at least through common hosting tools such as *Blogger*, rather than continuing to limit access only to those who had acquired the skills to author and upload individual home pages.

Exactly the same constraints, though with even greater capital costs, confronted those wishing to make "broadcast" video (television) and audio (radio) content available to audiences. In virtually every country, access to the broadcast spectrum is state-controlled, and licenses for access are extraordinarily expensive. Additional costs for studio time, on-air talent, the necessary audiovisual equipment, and other required infrastructure further compound the cost of production. The advent of audio and video blogging is a minor revolution from this point of view, as the cost of entry is minimal to the point of being trivial for the majority of citizens based in first-world nations and with disposable income. The technologies required for recording audiovisual materials are available domestically, and free audio and video editing software can be used to compile the work.

It is this ease of access to publishing, combined with the ease of distribution via a blog CMS, that offers the first major contribution of audio and video blogging to media culture. As a result, we have seen the rise of "citizen journalism" and alternative media in audiovisual formats as well as in the text-based forms examined elsewhere in this book, for example.[4] However, "alternative" in this context needs to be understood as it pertains to mainstream mass media, and certainly does not extend to alternative or other conceptions of audio and televisual media *per se*. While the viewpoints expressed in such content may provide an alternative to the views covered in the mainstream media, the forms of audio and video that are distributed and published via blogs remain resolutely conservative in their interpretation of audio and video as a material practice and object: by and large, they continue to adhere to those media forms and formats that have been established over the course of decades in the broadcast media.

Furthermore, it is possible to critique much of the recent commentary surrounding these alternative media practices (much of it appearing within blogs) in terms of a particular North American and, more specifically, U.S. experience of mass media which is marked by its homogeneity and commercial imperatives. (Most of the rest of the world has the experience of state-run media institutions, which generally support significantly more cultural and aesthetic diversity than do the mainstream U.S. media, and do not necessarily rely on advertising income.[5]) In addition, the "long tail" notwithstanding, most of the rhetoric about alternative-to-mass-media media practices is focused on a praxis of audience maximization that mimics the mass media. Regardless of whether alternative operators are attempting or claiming to offer diverse content, evaluation of the effectiveness of their operations is still premised on associating *influence* with *audience scale*.[6]

This does mean that audio and video blogging is a media practice that produces an interesting and potentially productive tension with existing audiovisual media institutions. It currently favors individual production over existing capital- and time-intensive industrial production models, supports a diversity of voices, and is comfortable with a range of genres and production standards. However, as those familiar with the histories of film, video, and sound will appreciate, such a list offers little, if anything, that distinguishes audio and video blogging from existing practices; there already is a strong and established tradition in each of these media that recognizes and supports an extremely diverse range of genres, production standards, and the legitimacy of self-defined creative constraints. What remains novel in the audio and video blogging model is only the range and ease of distribution.

What's Missing in Audio and Video Blogging

In this context, the differences between text-based and audio and video blogging must be made clear to see how constrained existing rich media blogging is as a *blog-based* practice. Blogs do considerably more than provide ease of publication and distribution for a diversity of voices. For example, as indicated above, they support and have led to the development of emergent communities of practice through the provision of blogrolls, trackbacks, and similar services. These products of good blogging practice should not be thought of as adjuncts or supplements to blogging, but are integral to blogging as a *different* writing practice, a writing that has recognized the network as an integral site of intensive connections. Blogs are about these relations between parts: while, say, printed books may exist in isolation, independently of one another, it is absurd to think of there being a single, isolated blog, precisely because a blog is determined by its relationship with other products of blogging, whether individual posts or entire blogs. If a blog were published in print (that is, published as a book), then it is no longer a blog; its "blogness" is broken.[7]

In the case of audio and video blogging, at present it is simply the presence of audio and video files that defines a blog as an audio or video blog. However, it is possible to remove the audio and video content from the context of the blog, and to publish it in other media, without causing an intrinsic change, or loss, to the material. Generally, it is possible to place the video content of a videoblog onto DVD and project it in a gallery or cinema, and yet the product is to all intents and purposes the same content as appears in the videoblog. Exactly the same process applies to audio content. This is why podcasting can be successful—there is nothing intrinsic in the media file that necessarily relates it to its "blogness," and so it survives this translation with ease. Indeed, it is feasible to imagine a television show for broadcast along the lines of "Australia's best videoblogs," and similarly a radio show based on "America's funniest podcasts."[8] Such thought experiments clearly indicate that, at this stage, audio and video blogs (or more precisely their RSS feeds, which feature rich media enclosures) are little more than distribution mechanisms for audiovisual content.

Toward New Forms of Audio and Video Blogging

But it is also possible to conceive of an alternative audio and video blogging practice that advances beyond such mere distribution of audiovisual content in the same way that text blogging as an alternative media form to traditional print provides features not available in the older medium. This alternative promotes rich media blogging at a level of sophistication that is far more developed than the manner in which audio and video blogging is currently practiced, through recognition of the formal material properties and affordances of the network that have been fundamental to the development and construction of *blogging* as a different writing practice. In analogy to the differences between printed and blogged text, then, the problem for audio and video blogging becomes one of envisaging how rich media artifacts may weave into, and be interlinked by, the network.

We have seen that podcasting and video blogging share some of the qualities of text blogging as they each enable multiple genres, diverse voices, and, for want of a better term, variable production standards. In each case, work ranges from the genuinely naïve, passing through would-be broadcast quality, through to a deliberately low-bit networked aesthetic. However, we should now also consider some of the elements still missing from audio and video blogging (in relation to blogging more generally) to see how else it *could* be different from other audiovisual forms—after all, the point of this form of blogging must ultimately be one of engaging in audio and video *blogging*, and not merely in providing audio and video on demand, or through syndication.

Currently, audio and video content in blogs is not usable in the ways that we take for granted with text; more specifically, audio and video blogging in its present forms is unable to manage most of the now-ordinary modes of engagement with posts in a text-based blog. For example, within any contemporary Web browser or RSS reader, it is possible to click'n'drag over text in a blog entry in order to copy this text using the software's generic "edit-copy" command. This text can then be pasted elsewhere, making it technically simple for others to quote, comment upon, or otherwise engage with the blogger's writing.

However, the same does not apply in the case of a communicative exchange between audio or video blogs: when listening to audio or viewing video from a rich media Weblog in a browser or RSS client, there is no similarly easy manner by which it is possible to select audio or video pieces and then paste them into a responding audio or video entry. If the audio or video file is opened in a specific player application (for example, QuickTime Player Pro), the viewer *can* copy and paste someone else's content into their own, in the manner possible with text-based posts—however, to do this they need to know considerably more about HTML, the Web, and file formats than is required of a user simply wishing to copy and paste what they find in a text blog. Why is this the case? Why, for example, does the QuickTime plugin not allow the user to nominate a passage of audio or video from an existing source, to copy and paste into a new piece of content directly from the browser's "edit" menu? This is exactly what is possible with the use of QuickTime Player Pro outside of the browser, and could easily be implemented into rich media blogging software. (One possible answer to this question may be a fear of legal implications on the side of the software manufacturer, of course—but as the following chapter by Fitzgerald and O'Brien shows, while illegal, unethical, or immoral uses of such a function are no doubt possible, there are also a great deal of uses, such as news reporting, commentary, parody, or artistic engagement, that clearly fall under applicable free speech and fair use provisions.)

In addition, the development of tools to edit and paste audio and video from within a browser or RSS client will raise a further anomaly in relation to audio and video in blogs: if a blogger quotes text in a blog post and follows the usual citational protocols of linking to the source of the quote in an individual entry, then the owner of the source blog will be alerted to the fact that a third party has written about their entry through referrer logs and/or the use of trackbacks. So far, this is not the case with audio and video: even if a blogger were able to open a third-party audio file in QuickTime Player, select a section of that file, copy and paste it into an audio blog post of their own, and publish that post, there is no equivalent to trackback available in an audiovisual context. Thus, the source of the third-party audio content will never know that their work has been "quoted" in this manner. But we should note that this is not primarily a technical question: an architecture as sophisticated as Quick-Time can already read XML and supports the dynamic editing of text tracks, so that it already offers almost all the functionality required to provide a time-based equivalent to trackback for audiovisual content. The absence of such functionality must therefore be seen as being caused by legal, theoretical, critical, or ideological concerns—perhaps among audio and video bloggers as much as among software developers. Indeed, the complete lack of interest in addressing the need for such technology in software development would seem to demonstrate the extent to which audio and video blogging as a practice looks backward to existing media for its methods, rather than toward the possibilities of blogging.

This simple example of quoting is useful to foreground the manner in which to date the key characteristic of audio and video blogging is merely the presence of audio and video content, and it demonstrates a continuing ignorance of the network and its affordances even among practitioners. This is evidenced not only in the simple problem of quotation, but also in the rise of syndication as a major component of audio and video blogging, resulting in the fact that media files are routinely viewed or used in isolation, with a dramatic loss of their networked and blogged contexts. In other words, when accessed in this way, the title and date of the entry, descriptive or associated text in the blog entry that accompanies the audio or video file, links within that text, and any comments or trackbacks attached to it, are lost to the viewer. The media file remains utterly mute in relation to the network, and so remains firmly embedded within the paradigms of audio and video as they have been traditionally conceived.

At the same time, it is already possible, using existing technologies, to include within audio and video files links that can be time-based and so only present during relevant periods of the entry, and in the case of video (or audio with a simple image track—for example, a still image), they can also be located on parts of the image just as in a traditional image map.[9] Once again, then, the problem is not technical: QuickTime and other media formats do offer these affordances, but at present the ability to link to and from sections of time-based media in the manner established by text blogs continues to fall outside of the paradigms by which rich media are understood, so that such tools are rarely utilized. The ultimate result is that, by comparison with text-based blogging, the existing uses of audio and video in blogs are far closer to print and the book than to the hypertextual fluidity of text within any garden-variety

Weblog. Once an audio or video blog entry of the current standard has been published (regardless of the efforts to produce it), it becomes a closed and whole object that is deaf to the network that it ostensibly participates within.

This begs the question of what it might mean for audio- and video-based media to be porous to the network—to allow quotation and interlinking and to develop a media form that is as permeable and granular as networked text. Such questions cannot be answered until tools are available that enable this to happen as easily as it does for text already. The narratives that could then be sung remain to be discovered. Blogs are the first online popular media form to recognize that relations between parts are a crucial quality for a properly net-worked practice—and while audio and video content remains closed to the net-work, audio and video "blogging" is limited to the placement of audio and video content *within* a blog, rather than true audio and video *blogging*.

Until rich media blogging integrates full network functionality into its de-velopment process, the traditional culture of the broadcast media remains un-contested and central to audio and video blogging. This accounts for why much of current audio and video blogging content mimetically mirrors the di-rect address forms popularized by mass media. This paucity of invention mis-takes media style for new paradigms, and with the rise of mobile non-networked devices, there still remains every opportunity for television and ra-dio to kill the yet-to-be-born audio and video blogging star.

NOTES

1. See Donald A. Norman, *The Design of Everyday Things* (London: MIT P, 1988).

2. This leaves to one side the more intriguing question of whether the tools appeared as a response to an emerging writing practice, or have in fact facilitated the development of a changed writing practice.

3. These qualities were celebrated by first-generation hypertext theorists eager to demonstrate how hypertext could facilitate broad "poststructural" revisions of writing and print. See, for example, Jay David Bolter, *Writing Space: The Computer, Hypertext, and the History of Writing* (Hillsdale, NJ: Lawrence Erlbaum, 1991); George P. Landow, *Hypertext 2.0: The Convergence of Contemporary Critical Theory and Technology* (Baltimore: Johns Hopkins UP, 1997). For a discussion of "intertwingling" and the docuverse, see Theodor Holm Nelson, *Literary Machines 91.1: The Report on, and of, Project Xanadu Concerning Word Processing, Electronic Publishing, Hypertext, Thinkertoys, Tomorrow's Intellectual Revolution, and Certain Other Topics Including Knowledge, Education and Freedom* (Sausalito, CA: Mindful Press, 1992).

4. See also Jen Simmons's introductory list at http://teaching.jensimmons.com/multimedia/2005/10/citizen-journalism.htm.

5. This is deserving of more analysis than I can provide here. Obviously, in the case of the United States, there is PBS and a strong tradition of public-access television, not to mention the presence of major Spanish-language broadcasting media. My point is that much of the celebration of pod- and videocasting as an alternative media practice is coming from U.S. commentators and is premised on hypostatizing U.S. mass media as hegemonic, yet does not acknowledge the existing presence of alternative media within the United States.

6. This is media studies 101—while many more people may read *News of the World*, *The Wall Street Journal* is almost certainly a much more influential newspaper. What matters is who reads it, and in the case of *The Wall Street Journal*, most, if not all, of the key economic players in the world are in its readership.

7. Here I am taking a formalist rather than essentialist point of view.

8. Indeed, the ease and rapidity with which traditional public broadcasters have been able to adopt podcasting demonstrates perhaps how little the format affects existing (old) media practices.

9. For a more detailed discussion of this, see Adrian Miles, "Media Rich versus Rich Media (or Why Video in a Blog Is Not the Same as a Video Blog)," *Blogtalk Downunder*, Sydney 2005, http://incsub.org/blogtalk/?page_id=74 (accessed 16 Nov. 2005). For examples see Adrian Miles, "BlogTalk Prototype 1," *Vlog 3.0*, 22 Apr. 2005, http://hypertext.rmit.edu.au/vlog/archives/2005/04/22/blogtalk-prototype-1/ (accessed 4 Nov. 2005), and Adrian Miles, "BlogTalk Prototype 3 (Small)," *Vlog 3.0*, 4 May 2005, http://hypertext.rmit.edu.au/vlog/archives/2005/05/04/blogtalk-prototype-3-small/ (accessed 4 Nov. 2005).

CHAPTER TWENTY

Bloggers and the Law

Brian Fitzgerald & Damien O'Brien

L ike any other communicative activity, blogging is regulated by the law.[1] It is therefore essential that bloggers be aware of the relevant laws relating to intellectual property (particularly copyright and trademark law), defamation, privacy, media and journalism, and employment relations.[2] This chapter will highlight some of the most important issues that can and have arisen in relation to blogs and the law. It goes without saying that law will provide a challenge to and be challenged by the distributed, original, and dynamic culture inherent in blogging.

Copyright Principles

Dating back to the Berne Convention on Literary and Artistic Works of 1886, there have been a variety of international treaties covering copyright. The multilateral Agreement on Trade Related Aspects of Intellectual Property Rights 1994 (TRIPS), the WIPO Copyright Treaty 1996 (WCT), the WIPO Performers and Phonograms Treaty 1996 (WPPT), and bilateral free trade agreements such as the Australia-U.S. Free Trade Agreement of 2004 are also part of this landscape. These treaties, which seek to harmonize copyright law across the world, tend to be implemented through national or domestic law— such as the Copyright Act or Code in each country. TRIPS requires member countries to enact copyright laws that uphold the Berne Convention and to ensure that adequate enforcement mechanisms are in place. Failure to do this can lead to the non-compliant member being taken to the World Trade Organization's (WTO) Dispute Settlement Body (DSB) and to trade sanctions being imposed.

The Berne Convention sets up a system of copyright law where the creator or author is the copyright owner at the outset, but they can (and often do) assign their copyright to a commercializing agent, such as a publisher, who then becomes the copyright owner. Copyright attaches to subject matter such as lit-

Moral Rights: independent of copyright, the right of an author to be attributed *as* author of copyright material, and to have the integrity of their work preserved.

erary, musical, dramatic and artistic works (including photographs), sound recordings, and film. A song, for instance, can be made up of a literary work (lyrics), musical work (score), and a sound recording, with the different aspects owned by different copyright owners. In most countries except the United States (where such rights in copyright law are almost nonexistent), creators or authors hold moral rights to be attributed as the author of copyright material and to ensure the integrity of their work (not to have it mutilated in such a manner as to cause dishonor). These rights are not usually assignable, but in some countries they can be waived or overridden with consent. In countries such as France and Germany, waiver is not possible.

Moral rights stay with the creator or author. Economic rights go with the copyright owner. The key economic rights give the copyright owner power to control things such as reproduction and communication to the public. If any of these rights are exercised, the permission of the copyright owner will normally be required unless one of the exceptions applies (under the various copyright laws) that allow use of copyright material without the permission of the copyright owner. In the United States, the fair use provision allows a

Fair Use / Fair Dealing: provisions that allow the use of copyright material in specific contexts, often including news reporting, scholarly use, critique, or parody.

broad range of uses without permission for purposes such as parody or critique, while in other countries narrower notions such as fair dealing allow strictly controlled use for research, news reporting, review, or criticism. There may also be statutory licenses (e.g., for private use) that allow use of the material without the permission of the copyright owner if compensation is paid.

Copyright Issues

Copyright raises numerous legal issues for blogs, particularly in relation to the reproduction and communication to the public of any text, images, or sound recordings.[3] While copyright law does not protect ideas, only the "expression" of ideas (the way things are written), it will have broad applicability to much of the material that appears on blogs. The initial questions will be:

1. Is the material appearing on a blog copyright material?
2. If yes, is permission to use it required from the copyright owner, or does copyright law provide the permission such as through fair use or fair dealing?
3. If remixing or endorsement is involved, are moral rights an issue?

Text

Generally, written commentary supplied by a host blogger or contributors to a blog will be classified as a literary work owned by the creator, unless it has been assigned. It should also be noted that material created through the course of employment will normally be owned by the employer. Therefore, any presentation of textual material on a blog will most likely involve reproduction and communication to the public, and as such the permission of the appropriate copyright owner must be obtained or a legal right to use without permission (e.g., fair use or fair dealing or statutory license) must be able to be asserted. A failure to obtain or be able to assert such permission, either by mistake or deliberately, will be likely to infringe the owner's copyright in the material.[4] Some blogs are now using Creative Commons (CC)-style licenses, which allow people to take material and reutilize it on certain conditions. CC licenses act like a permission in advance and allow further negotiability of the written commentary. However, bloggers should be aware that a failure to comply with the relevant CC license may still be an infringement of the copyright owner's work.[5]

> **Creative Commons:** the suite of Creative Commons licenses builds upon the "all rights reserved" of traditional copyright to create voluntary "some rights reserved" copyright regimes.

Images

Graphics in the form of artistic drawings and photographs used to enhance the design, ambience, or the substantive discussion of the blog also need to be assessed in terms of copyright law. If they have been taken from another site, a blogger will need to ensure that they have permission to use them, either express or under some legal doctrine such as fair use. Without such permission, the reproduction or communication to the public of the images will amount to an infringement of copyright.[6]

In most copyright regimes, the taking of an "insubstantial" amount of copyright material, which usually relates to quality not quantity (but could relate to both in some countries), will not amount to copyright infringement. As well, in all categories the issue of whether the representation on the blog infringes moral rights, especially those of attributing the author/creator and of maintaining the integrity of the material (although in some countries, such as France, other moral rights will apply), will need to be considered.

Music

More specifically, in relation to music, those blogs that incorporate MP3s, commonly known as MP3 or audio blogs, should be aware of the potential of copyright infringement from the unauthorized reproduction and communication to the public of songs. Unless permission has been given by the copyright owner, a license has been obtained from a performing rights collecting society to play the song via the Website, or fair use/dealing applies, the reproduction or communication to the public is likely to be unauthorized.

Since their inception in early 2003, MP3 blogs have become a popular way for people to share their favorite music in the digital environment. The concept of an MP3 blog essentially involves the combination of an online journal with a music column that features MP3 music files that are available for download. Generally, MP3 blogs contain one or two tracks from a CD album that are available for download. This is usually accompanied by the traditional blog, which features a commentary on or review of the track and the artist. Readers are then encouraged to download the music, read the accompanying review, and share their thoughts online. The MP3 files that are contained on the blogs are either available for download directly from the blog itself, or through a link to another site where the MP3 files have been uploaded. However, in most cases, the MP3 files are only available to download for a couple of days. By their very nature, most such MP3 blogs tend to feature obscure "musical nuggets," such as hard-to-find and often outdated tracks that are restricted to a particular musical sub-genre or theme. MP3 blogs tend to fall into two categories: those that provide music with the copyright owner's permission and those that do not. It is the latter which will have implications for copyright law.

Thus far, MP3 blogs have managed to avoid the wrath of the music industry and are therefore yet to be legally challenged.[7] However, it has been well documented that they exist within a so-called legal grey area, and it may only be a matter of time before the law turns its attention to MP3 blogs. Recently the Recording Industry Association of America (RIAA) stated that in terms of piracy, MP3 blogs are an issue that it is closely monitoring, and that at any time it could decide to make enforcement a priority. The main reason for the survival of MP3 blogs to date is their relatively low profile, with even the most popular MP3 blogs having only a few thousand regular visitors. This is a far cry from the millions of people who engage in peer-to-peer file sharing through programs like WinMX or Kazaa. In addition, most MP3 blogs tend to feature music that is no longer termed as mainstream and has often been out of the public eye for a long time.

However, in spite of these factors, while MP3 blogs continue to feature tracks without the permission of the copyright owner, they run the risk that they will infringe copyright law. Under copyright law, bloggers will infringe copyright when they perform any of the acts within the copyright owner's exclusive rights and cannot avail themselves of an exception such as fair use or fair dealing. In the context of a sound recording, this will most often occur on

MP3 blogs where the host blogger makes a copy of the sound recording or where they communicate the recording to the public by posting it to the blog. In this scenario there will also most likely be a copyright infringement of the musical and literary work, as well as of the sound recording. In light of recent decisions such as *Universal Music Australia Pty Ltd v. Cooper*,[8] host bloggers also need to be mindful of contributory or authorization liability for facilitating copyright infringement through hypertext linking.

Assuming an action for copyright infringement can be made out against an MP3 blog, an issue that arises is whether MP3 blogs can avoid liability through the notions of fair use in the United States and fair dealing in other countries. Under the fair use doctrine, users of copyright material can reproduce or otherwise infringe copyright for purposes such as criticism, comment, news reporting, teaching (including multiple copies for classroom use), scholarship, or research. However, this needs to be assessed in light of the following criteria: "the purpose and character of the use including whether the use is of a commercial nature or for a non profit educational purpose," "the nature of the copyrighted work," "the amount and substantiality of the portion used in relation to the copyrighted work as a whole," and the "effect of the use upon the potential market for or value of the copyrighted work."[9]

In countries such as Australia, Canada, and the United Kingdom (and any other country employing the notion of fair dealing), the key issue will be whether MP3 blogs come within the fair dealing defense of criticism or review. Under this notion, a musical or literary work or a sound recording may be fairly dealt with, without infringing copyright, for the purposes of criticism or review. In *Warner Entertainment Co Ltd v. Channel 4 Television Corp PLC*, Lord Justice Henry stated that the question to be answered in assessing whether a dealing is fair or not is "is the [work] incorporating the infringing material a genuine piece of criticism or review, or is it something else, such as an attempt to dress up the infringement of another's copyright in the guise of criticism."[10]

From this arises the issue of whether the commentary and review posted on MP3 blogs will be sufficient to constitute criticism and review under the notion of fair dealing. Given the differing nature of each MP3 blog, it is not possible to provide one universally applicable answer; rather, each site will need to be assessed on a case-by-case basis. However, it is possible to identify a number of key indicators that may suggest whether the fair dealing defense of criticism or review will be applicable in a given case. The primary determining factor will be the amount of commentary that is featured on the MP3 blog itself. In the case where an MP3 blog contains quite detailed commentary, a court may be inclined to view it as a genuine piece of criticism or review. This is to be distinguished from those sites that do not contain detailed commentary and are likely to be viewed as an infringement of copyright. Another determining factor will be the number of tracks that are available for download on the MP3 blog. Where there are only one or two tracks available, a court may be more willing to allow the criticism or review defense. However, MP3 blogs that contain an entire album or a substantial number of tracks will most likely not be afforded the defense of fair dealing. In summary, it would appear that as a general guide, where an MP3 blog is on its face nothing more than an attempt to disguise copyright infringement, the defense of fair dealing will not be allowed. However, if the MP3 blog is a genuine piece of criticism or review, and is on a small scale, then a court may be inclined to allow the fair dealing defense.

Trademarks

A trademark is a sign or symbol used to identify the source of goods or services. For example, "Coke" identifies a particular brand of cola soft drink made by a particular company. Customers can rely on that branding whether they are in Beijing, Tokyo, Delhi, New York, or Paris to know what they are drinking and what it will taste like. As a general rule, persons cannot use another's trademark on similar goods and services so as to create consumer confusion, nor can they (in many countries) dilute through use the value of well-known or famous trademarks by placing them on any type of goods or services.

Culture Jamming: using the tools of advertising (including the usually unauthorized use of familiar corporate designs and trademarks) for the promotion of oppositional messages.

If a blog is reproducing a trademark to label similar goods and services, then permission to use will be needed unless the blogger is engaging in comparative advertising. If the trademark is being used to criticize or culture-jam a famous trademark, then the issue of trademark parody needs to be considered. As use of the trademark in this manner is not "use of the trademark as a trademark," the critical question will be to what extent dilution-type laws allow this kind of activity. The U.S. courts have tended to allow trademarks to be reproduced on goods and even sold so long as it is a "take off" and not a "rip off."[11] The introduction of a federal trademark dilution law has brought some uncertainty in the case law as to the legality of parody, yet there seems to be a clear argument that "non-commercial speech" (in essence social commentary) involving a trademark is protected by the First Amendment, and that such use will not amount to dilution.[12] The critical question will be whether parody devalues the mark—and if the answer is yes, a further question will be whether the parody devalues the mark in its ability to draw consumers, or only within a broader social consciousness.[13]

Trademarks can also be an issue in relation to domain names. Using domain names that create consumer confusion or dilution may lead to liability. A key issue in relation to copyright and trademark liability will be the extent to which a blog can be liable for merely *linking* to an infringing Websitē.[14] A systematic form of linking that brings popularity and/or revenue is more likely to cause legal problems; however, any infringement can technically establish liability.[15]

Internet Jurisdiction and Defamation

Much of what has been discussed above assumes that the laws mentioned are enforceable. However, commentators frequently ask how anyone can enforce any law covering Internet activities.[16] In this context, it is important to recognize that at the end of the day, if a person or company can convince a court that it has jurisdiction over a matter, and can then manage to win that case and enforce it against the defendant's assets, the law will be enforced. In this way, enforcement can happen even in an Internet context. The critical step is convincing a court that it has jurisdiction, and then obtaining and enforcing judgment in whatever country the defendant's assets are located. That last step can be problematic when trying to enforce, by registering the judgment in the United States, laws that infringe the First Amendment to the U.S. Constitution.

Internet Jurisdiction

Jurisdiction refers to the ability of a state to legislate, administer, and enforce legal sanctions.[17] Under international law, a state has the right to exercise its jurisdiction on the bases of sovereignty over the territory, nationality of the perpetrator or victim, protection or security, and the universal nature of the crime. However, any exercise of jurisdiction must be undertaken with a degree of reasonableness.

The traditional idea of a state exercising sovereignty over individuals within a defined territory is fundamentally challenged by the transnational nature of the Internet. At one extreme is the view that jurisdiction will be established in every country where one can view or access a Website; there have been many cases in the United States testing this proposition.

Two approaches have emerged from the case law:

1. the *Zippo* sliding scale approach
2. the *Calder* "effects" and "targeting" approach

In *Zippo* the court explained that a finding of jurisdiction was contingent upon the nature of the Website and sought to employ a sliding scale test.[18] A fully interactive Website would establish jurisdiction, while a passive Website used for mere advertising would not. In principle, to establish jurisdiction, the Website has to reach out and touch the territory in question.

U.S. courts have also utilized the *Calder* "effects" test to establish jurisdiction. In essence,

1. where an act is done intentionally,
2. has an effect within the forum state (the state in which the litigation is occurring), and
3. is directed or targeted at the forum state,

then jurisdiction will be satisfied. This approach is evidenced in the recent case of *MGM v. Grokster*,[19] where a California court assumed jurisdiction in a case relating to copyright infringement. One of the defendants in this case distributed through a Website a software product known as Kazaa Media Desktop (KMD), which was used to share digital entertainment such as music and film. The court held that jurisdiction was established on the basis that the software had an impact or effect in California, as it was the movie capital of the world, and that the software had been targeted at California.[20]

A recent judgment of the High Court of Australia in relation to defamation suggests that jurisdiction may be established on a broader basis than an interactive or targeting Website.[21] In *Dow Jones & Company Inc v. Gutnick*, content that was created in New York, loaded on a server in New Jersey, and available for access through the Internet in the state of Victoria in Australia, was held to be actionable in the courts of Victoria. The court was asked:

1. whether jurisdiction was established,
2. whether Victorian law applied, and
3. whether Victoria was a clearly inappropriate forum (the *forum non conveniens* question).

The court in effect disposed of three issues in one blow by holding that "publication" for the purpose of defamation occurred where the defamatory material was comprehended. It explained:

> In defamation, the same considerations that require rejection of locating the tort by reference only to the publisher's conduct, lead to the conclusion that, ordinarily, defamation is to be located at the place where the damage to reputation occurs. Ordinarily that will be where the material which is alleged to be defamatory is available in comprehensible form assuming, of course, that the person defamed has in that place a reputation which is thereby damaged. It is only when the material is in comprehensible form that the damage to reputation is done and it is damage to reputation which is the principal focus of defamation, not any quality of the defendant's conduct. In the case of material on the World Wide Web, it is not available in comprehensible form until downloaded on to the computer of a person

who has used a web browser to pull the material from the web server. It is where that person downloads the material that the damage to reputation may be done. Ordinarily then, that will be the place where the tort of defamation is committed.[22]

Interestingly, only days later a U.S. court came to the opposite conclusion on very similar facts, holding that unless the Website containing the defamatory statement was targeted toward the forum state, jurisdiction would not be established.[23]

Defamation

The publication of defamatory material will be an actionable wrong in most jurisdictions throughout the world and may subject a person to damages or injunctive relief. Defamation generally consists of the communication of a defamatory statement concerning the plaintiff (that is, the person suing) to a person other than the plaintiff. In relation to blogs, this communication will most likely occur through electronic text, images, and sound. If such communication becomes known to another person, and amounts to a publication of a defamatory statement, then a person will have, subject to defenses such as truth and public interest, a legal action for defamation.[24]

The essential elements of any legal action for defamation are:

1. a defamatory imputation or statement, which
2. refers to the plaintiff (the person suing), and
3. is published to a third party.

In relation to blogs and defamation, a recent Australian decision of the Supreme Court of New South Wales appears to confirm the view that the courts will treat blogs in the same way as if the defamatory material had been communicated in the non-digital, paper-based world. In *Kaplan v. Go Daddy Group Inc*,[25] the (second) defendant created a blog, under a disparaging domain name, which encouraged other readers to post derogatory comments about the plaintiff's business. The effect was that six postings were created, all of which contained defamatory comments about the plaintiff's business. The plaintiffs applied to the court for an interim injunction to prevent the (second) defendant from maintaining the blog. In granting the injunction, Justice White held that there was a *prima facie* case that the (second) defendant had committed the tort of injurious falsehood (that is, a false assertion that is intended to injure a person's property, product, or business), as the (second) defendant had sought to use the blog to publish false and misleading disparagements of the plaintiff.

It is also important to point out that U.S. defamation law, informed by the First Amendment,[26] will allow a broader scope to make comments about public figures than will the laws of other countries. Under U.S. law, a public figure will generally be held to be someone "who has actively sought, in a given matter of public interest, to influence the resolution of the matter."[27] Obvious public figures include government employees, senators, and presidential candidates. However, there are also "limited-purpose public figures"; these are generally people who "voluntarily participate in a discussion about a public controversy, and have access to the media to get his or her own view across."[28] In Australia, this notion is limited to persons having a relationship to the workings of government.[29]

The more a person has control over the material uploaded to a blog, the more likely they will be seen to be an editor. In *Cubby v. Compuserve*,[30] a U.S. court held that an Internet Service Provider (ISP) was not liable for defamatory material housed on its server so long as it did not have knowledge of the existence of the infringing material, or place it there. The court explained that an ISP was an innocent distributor and that, much like a person selling a newspaper on a street corner, or a library, it should not be held liable. In the later U.S. case of *Stratton Oakmont v. Prodigy*,[31] the court suggested that editorial control on any level could lead to the finding that an ISP had moved from being an innocent distributor to being a publisher. In this case, Prodigy was seen to be engaging in editorial control by offering a "clean service," and thereby was held liable for defamation. This decision meant that the more an ISP moved to clean up its server, the more likely it was to be held liable as a publisher. This led to intense lobbying in the United States and the passage of the Good Samaritan statute—section 230 of the *Communications Decency Act* (1996) (CDA)—which immunizes ISPs from liability for defamation in certain circumstances. Whether a host blogger can claim immunity under the CDA for comments uploaded by contributors is yet to be determined.[32]

Employee Blogging

One of the important legal issues relating to blogs is to what extent an employer will permit an employee to blog about work-related matters.[33] One well-known example of this was Ellen Simonetti, a Delta Airlines flight attendant whose employment was terminated for allegedly posting images of herself in uniform, which infringed on the Delta Airlines uniform policy.[34] As a result of this case and many other similar cases, employers are now considering including specific blogging provisions in employment contracts. These clauses are an attempt to stop employees from referring to their employers in their personal blogs. Some employers have even taken steps to ensure that employment contracts disallow employees from blogging at all. Therefore, bloggers must be aware of what their employment contract states in regards to employee blogging. They should be aware that a failure to comply with any contractual provisions may enable their employer to terminate their employment for a breach of contract.

Journalistic Privilege

Under U.S. law, most state and federal constitutions provide for a journalistic privilege to allow journalists to keep the names and unpublished information of their sources confidential. The question that arises in relation to blogs, then, is to what extent a blogger can be classified as a journalist, and thus afforded the protection of journalistic privilege. In the recent Californian decision of *Apple Computer Inc v. Does*,[35] a Santa Clara court ruled that a blogger's Internet Service Provider (in this case Nfox) could be required to reveal to attorneys from Apple Computer Inc. the identities of people using an email service to the *Powerpage* blog. This case related to the posting on the *Powerpage* blog of an exact copy of the drawing of the "Asteroid" created by Apple, and of other technical specifications taken from a confidential slide set.[36] The court held that regardless of whether the bloggers who intervened in the case were journalists or not, "shield laws" protecting sources could not trump trade secret laws in this instance.

It is also interesting to note that in the recent case of *Doe v. Cahill*[37] before the Supreme Court of Delaware, it was held that a strict requirement showing the strength of the case needed to be met before the identity of a blogger could be unmasked in the process of litigation for defamation.

Immunity of the Host Blogger for Infringement?

Host bloggers need to be mindful that accessory liability for authorizing, facilitating, or contributing to unlawful infringement may establish liability. This has been a key issue in recent cases in the copyright area relating to peer-to-peer file sharing software developers such as Napster, Grokster, and Kazaa.

In a number of countries, most notably the United States, through section 512 of the *Digital Millennium Copyright Act 1998* (DMCA), online service providers are allowed a degree of immunity from liability for copyright infringement in certain circumstances. This type of provision requires a number of conditions, including that the service provider register under the DMCA, automation of the uploading, and lack of knowledge of the infringing material, and allows for take-down notices to be sent to the service provider that must be complied with until a counter-notice is supplied. A key question will be whether the host blogger qualifies for immunity under this type of provision. At present it is doubtful that they will, due to the definition of "online

service provider" and the myriad formal requirements that need to be satisfied.[38]

Conclusion

Blogs provide an exciting new way in which to manage and present information and arguments about the world around us. Their informality and
spontaneity leads lawyers to caution that while a rhetoric of free expression
rules the Internet, the reality is that corporate and personal interests in the
form of intellectual property and reputation will continue to be a regulating
factor. Where a blog is popular or generates economic revenue (directly or indirectly), aggrieved parties will be motivated (more than usual) to sue.

NOTES

1. This article is not provided as legal advice and should not be relied on as such.

2. See, generally, Electronic Frontier Foundation, "Legal Guide for Bloggers," 6 July 2005, http://www.eff.org/bloggers/lg/ (accessed 18 Nov. 2005).

3. See Anne Fitzgerald and Brian Fitzgerald, *Intellectual Property in Principle* (Sydney: Thomson, 2004).

4. For example, see the *Washington Post* "cease and desist" notice relating to the unauthorized reproduction of a newspaper article, in "Washington Post Doesn't Mean 'Re-post' on Blog," *Chilling Effects*, 2005, http://www.chillingeffects.org/dmca512/notice.cgi?NoticeID=2073 (accessed 18 Nov. 2005); for an Australian example, see Hobsons Australia's "cease and desist" notice in regard to the reproduction of the *Good Universities Guide*, in "Hobsons Schools Blogger for Copying 'Good Universities Guide,'" *Chilling Effects*, 2005, http://www.chillingeffects.org/dmca512/notice.cgi?NoticeID=2177 (accessed 18 Nov. 2005).

5. For example, see "Author Complains of Copying in Blog," *Chilling Effects* 2005, http://www.chillingeffects.org/dmca512/notice.cgi?NoticeID=2286 (accessed 18 Nov. 2005) in relation to a failure to comply with the "no derivative works" clause.

6. For example, see the Playboy "cease and desist" notice relating to the reproduction of images, in "Playboy Complains of Blog-Posted Images," *Chilling Effects*, 2005, http://www.chillingeffects.org/dmca512/notice.cgi?NoticeID=2490 (accessed 18 Nov. 2005).

7. See *Universal Music Australia Pty Ltd v. Cooper* 2005, FCA 972, http://www.austlii.edu.au/au/cases/cth/federal_ct/2005/972.html (accessed 18 Nov. 2005); *Universal Music Australia Pty Ltd v. Sharman License Holdings Ltd* 2005, FCA 1242, http://www.austlii.edu.au/au/cases/cth/federal_ct/2005/1242.html (accessed 18 Nov. 2005).

8. *Universal Music Australia Pty Ltd v. Cooper.*

9. Section 107 U.S. Copyright Act 1976.

10. *Warner Entertainment Co Ltd v. Channel 4 Television Corp PLC* 1993, 28 IPR 459, p. 468.

11. *Nike Inc v. "Just Did It" Enterprises* 6 F3d 1225, 7th Cir. 1993, pp. 1227–28; *The Coca Cola Co v. Gemini Rising Inc* 346 F. Supp. 1183, E.D.N.Y. 1972; *Anheuser-Busch Inc v. L & L Wings Inc* 962 F. 2d 316, 4th Cir. 1992; G. Mayers, "Trademark Parody: Lessons from the Copyright Decision in *Campbell v. Acuff-Rose Music Inc*," *Law & Contemporary Problems* 60.181 (1996).

12. See *Mattel Inc v. Walking Mountain Productions* 353 F. 3d. 792, 9th Cir. 2004; *Mattel Inc v. NCA Records Inc* 296 F 3d 894, 9th Cir. 2002, Cert. Denied 537 U.S. 1171, 2003; *Dr Seuss Enterprises v. Penguin Books USA* 109 F 3d 1394, 9th Cir. 1997; E. Gredley and S. Maniatis, "Parody: A Fatal Attraction? Part 2: Trade Mark Parodies," *European Intellectual Property Review* 8.412 (1997).

13. *British Columbia Automobile Assn v. Office and Professional Employees International Union Local 378*, 2001, B.C.J. No. 151 at 165-8; *Mattel Inc v. Walking Mountain Productions* 353 F. 3d. 792 at 812, 9th Cir. 2004; *Mattel Inc v. NCA Records Inc* 296 F 3d 894 at 902-7, 9th Cir. 2002, Cert. Denied 537 U.S. 1171, 2003.

14. *Ticketmaster Corp v. Tickets.com Inc* 54 USPQ. 2d 1344, C.D. Cal. 2000.

15. *Universal Music Australia Pty Ltd v. Cooper.*

16. See, generally, Brian Fitzgerald, "*Dow Jones & Company Inc v. Gutnick* [2002] HCA 56: Negotiating American Legal Hegemony in the Transnational World of Cyberspace," *Melbourne University Law Review* 27.590 (2003), http://www.austlii.edu.au/au/journals/MULR/2003/21.html (accessed 18 Nov. 2005); Brian Fitzgerald, Gaye Middleton, and Anne Fitzgerald, *Jurisdiction and the Internet* (Sydney: Thomson/LBC, 2004); Barry Sookman, *Computer, Internet and Electronic Commerce Law* (Toronto: Carswell Canada, 1991), Chapter 11.

17. Jurisdiction may be conceptualized as the consent of the governed to be brought before the courts in question (social contract rationale): see Lea Brilmayer, "Consent, Contract, and Territory," *Minnesota Law Review* 74.1 (1989); or more as an efficient mechanism for coordinating international litigation (coordination rationale): Berkman Center, *iTunes White Paper*, 2004, http://cyber.law.harvard.edu/media/uploads/81/iTunesWhitePaper0604.pdf, pp. 91ff.

18. *Zippo Manufacturing v. Zippo.com* 952 F. Supp 1119, W.D. Pa 1997.

19. *MGM Metro Goldwyn Mayer Studios Inc v. Grokster Limited*, 243 F Supp 2d 1073, CD Cal, 2003. See also *Young v. New Haven Advocate* 315 F.3d 256, 4th Cir 2003.

20. Cf. *Pavlovich v. DVD Copy Control Association* 127 Cal. Rptr 2d 329, Cal. 2002.

21. *Dow Jones & Company Inc v. Gutnick*, 2002, HCA 56, http://www.austlii.edu.au/au/cases/cth/HCA/2002/56.html (accessed 18 Nov. 2005). See also *Lewis v. King*, 2004, EWCA Civ 1329; *Bangoura v. Washington Post Co*, 2004, 235 D.L.R. (4th) 564.

22. *Dow Jones & Company Inc v. Gutnick*.

23. *Young v. New Haven Advocate*. See also *Griffis v. Luban* 646 N.W. 2d 527, S. Ct. Minn. 2002; *Yahoo! Inc v. La Ligue Contre le Racisme et l'Antisémitisme* 379 F. 3d 1120, 9th Cir. 2004.

24. See, for example, the "cease and desist" notice relating to allegedly libelous statements on a blog in "Allegedly Libelous Statements on Blog," *Chilling Effects*, 2005, http://www.chillingeffects.org/defamtion/notice.cgi?NoticeID=1901 (accessed 18 Nov. 2005).

25. *Kaplan v. Go Daddy Group Inc*, 2005, NSWSC 636 (White J), http://www.austlii.edu.au/cases/nsw/supreme_ct/2005/636.html (accessed 18 Nov. 2005).

26. *New York Times Co v. Sullivan* 376 U.S. 254, 1964.

27. Electronic Frontier Foundation, "Online Defamation Law," 2005, http://www.eff.org/bloggers/lg/faq-defamation.php (accessed 18 Nov. 2005).

28. Electronic Frontier Foundation, "Online Defamation Law."

29. *Lange v. Australian Broadcasting Corporation*, 1997, 189 CLR 520, http://www.austlii.edu.au/au/cases/cth/HCA/1997/25.html (accessed 18 Nov. 2005).

30. *Cubby v. Compuserve* 776 F Supp 135, SDNY 1991.

31. *Stratton Oakmont v. Prodigy*, 1995, NY Misc LEXIS 229; 1995 WL 323710, NY Sup Ct 24 May 1995.

32. See, further, *Zeran v. AOL* 129 F.3d 327, 4th Cir. 1997.

33. See, generally, Ken Ebanks, "Into the Blogosphere: Managing the Risks and Rewards of Employee Blogging," *The Computer and Internet Lawyer* 22.1 (Oct. 2005), p. 1.

34. See http://queenofsky.journalspace.com/.

35. *Apple Computer Inc v. Does*, D.S.C. 03/11/2005.

36. Current Developments, "Journalistic Privilege Did Not Protect Source of Trade Secret Posted by Blogger," *The Computer and Internet Lawyer* 22.6 (June 2005), p. 33.

37. *Doe v. Cahill*, 5 Oct. 2005, No. 266/2005, Supreme Court of Delaware. This case raises the issue as to whether and to what extent bloggers have a First Amendment right to speak anonymously.

38. See, further, Attiya Malik, "Are You Content with the Content? Intellectual Property Implications of Weblog Publishing," *The John Marshall Journal of Computer and Information Law* 21.439 (2003), pp. 487–88; *A&M Records Inc v. Napster Inc* 239 F.3d 1004, 9th Cir. 2001. See also "Gettyimages Sends 512 Notice to Blogger," *Chilling Effects*, 2005, http://www.chillingeffects.org/dmca512/notice.cgi?NoticeID=871 (accessed 18 Nov. 2005), relating to a "cease and desist" notice served on an ISP pursuant to s 512(c) Digital Millennium Copyright Act 1998 in relation to photographs posted on a blog.

CHAPTER TWENTY-ONE

Blogs and the
Communications Renaissance

Joanne Jacobs & Douglas Rushkoff

Finally it's becoming practical for people with little or no technical expertise whatsoever to find their voice.

Now we just need ways to make sure we know how to hear them....[1]

—Tom Coates (2003)

Finding people with similar interests to yourself can be a challenging experience. Finding out how to communicate effectively with people of similar interests has been an ongoing challenge. In spite of the promise of new technologies acting as an agora for political debate and cyberdemocracy in the fashion documented by Mark Poster,[2] the realities of bandwidth scarcity, the intellectual property issues associated with posting rich media content, and a user-oriented need for maximum communication with minimum time wastage have meant that the vast majority of communications between people online are text-based. And the nature of text-based communications is such that the practice of communicating ideas and cultures becomes a frustrating process of negotiating meaning and searching for synergy. This could well be a reason why blogs have been so successful at defining new cultures for the sharing of ideas. Dominated by text (in spite of the growth of moblogging, picture, and vidblogging—on the latter, see chapter nineteen by Adrian Miles in this volume), text-based blogs allow for conversations and cultures to develop in an almost evolutionary fashion, growing and changing in accordance with reader interests and influences. Thus it is blogs that come closest to achieving the vision for cyberspace first promulgated by advocates of technocracies—cyber-utopias for cultural expansion, political debate, and economic efficiency. Blogs provide us with the basis for developing the triple crown of "faster, cheaper, better"; information of interest to any specific community is available on the fly, at almost zero cost for all members of a community, and with more reliability and cross-referencing than was ever before available. In terms of modern cultural development, blogs may well be the source of the communication renaissance.

This chapter demonstrates the power of blogs as a means of constructing and negotiating ideas. Excerpts from blog posts by Douglas Rushkoff have been interspersed with commentary by *Uses of Blogs* co-editor Joanne Jacobs.

The Communications Renaissance

» Douglas Rushkoff — Renaissance *Now?* (13 June 2002)[3]

The birth of the Internet era was considered a revolution, by many. My best friends—particularly those in the 'counterculture'—saw in the Internet an opportunity to topple the storytellers who had dominated our politics, economics, society, and religion, in short, our very reality, and to replace their stories with ones of our own. It was a beautiful and exciting sentiment, but one as based in a particular narrative as any other. Revolutions simply replace one story with another. The capitalist narrative is replaced by the communist; the religious fundamentalist's for the agnostic's. The means may be different, but the rewards are the same, as is the exclusivity of their distribution. That's why they're called revolutions; we're just going in a circle.

I prefer to think of the proliferation of interactive media as an opportunity for renaissance: a moment when we have the opportunity to step out of the story, altogether. Renaissances are historical instances of widespread recontextualization. People in a variety of different arts, philosophies, and sciences have the ability to reframe their reality. Quite literally, renaissance means "rebirth." It is the rebirth of old ideas in a new context. A renaissance is a dimensional leap, when our perspective shifts so dramatically that our understanding of the oldest, most fundamental elements of existence changes.

Take a look back at what we think of as the original Renaissance—the one we were taught in school. What were the main leaps in perspective? Well, most obviously, perspective painting itself. Artists developed the technique of the "vanishing point" and with it the ability to paint three dimensional representations on two-dimensional surfaces. The character of this innovation is subtle, but distinct. It is not a technique for working in three dimensions; it is not that artists moved from working on canvas to working with clay. Rather, perspective painting allows an artist to relate between dimensions. It is a way of representing three-dimensional objects on a two-dimensional plane.

Likewise, calculus—another innovation developed not quite during, but immediately in the wake of Renaissance advances—is a mathematical system that allows us to derive one dimensional perspective from another. Most simply, it is a new way of describing curves with the language of lines, spheres with the language of curves, or speed in the language of distance. The leap from arithmetic to calculus was not just a leap in our ability to work with higher dimensional objects, but a leap in our ability to relate the objects of one dimension to the objects of another. It was a shift in perspective that allowed us to orient ourselves to mathematical objects from beyond the context of their own dimensionality.

The other main features of the Renaissance permitted similar shifts in perspective. Circumnavigation of the globe changed our relationship to the planet we live on and the maps we used to describe it. The maps still worked, of course—only they described a globe instead of a plane. Anyone hoping to navigate a course had to be able to relate a two-dimensional map to the new reality of a three-dimensional planet. Similarly, the invention of moveable type and the printing press changed the relationship of author and audience to text. The creation of a manuscript was no longer a one-pointed affair. Well, the creation of the first manuscript still was—but now it could be replicated and distributed to everyone. It was still one story, but now it was subject to a multiplicity of individual perspectives. This lattermost innovation, alone, changed the landscape of religion in the Western World. Individual interpretation of the Bible led to the collapse of Church authority and of the unilateral nature of its decrees. Everyone demanded his or her own relationship to the story.

In all these cases, people experienced a very particular shift in their relationship to and understanding of dimensions. Understood this way, a renaissance is a moment of reframing. We step out of the frame as it is currently defined, and see the whole picture in a new context. We can then play by new rules.

The great Renaissance was a simple leap in perspective. Instead of seeing everything in one dimension, we came to realize there was more than one dimension on which things were occurring. Even the Elizabethan world picture, with its concentric rings of authority—God, king, man, animals—reflects this newfound way of contending with the simultaneity of action of many dimensions at once.

The evidence of today's renaissance is at least as profound as that of the one that went before. The 16th century saw the successful circumnavigation of the globe via the seas. The 20th century saw the successful circumnavigation of the globe from space. The first pictures of earth from space changed our perspective on this sphere, forever. In the same century, our dominance over the planet was confirmed not just through our ability to travel around it, but to destroy it. The atomic bomb (itself the result of a rude dimensional interchange between submolecular particles) gave us the ability to destroy the globe. Now, instead of merely being able to circumnavigate "God's" creation, we could actively destroy it. This is a new perspective.

We also have our equivalent of perspective painting, in the invention of the holograph. The holograph allows us to represent not just three, but four dimensions on a two-dimensional plate. When the viewer walks past a holograph, she can observe the three-dimensional object over a course of time. A bird can flap its wings in a single picture. But, more importantly for our renaissance's purposes, the holographic plate itself embodies a new renaissance principle. When the plate is smashed into hundreds of pieces, we do not find that one piece contains the bird's wing, and another piece the bird's beak. No, each piece of the plate contains an image of the entire subject, albeit a faint one. When the pieces are put together, the image achieves greater resolution. But each piece contains a representation of the totality—a leap in dimensional understanding that is now informing disciplines as diverse as brain anatomy and computer programming.

Our analog to calculus is the development of systems theory, chaos math, and the much-celebrated fractal. Confronting non-linear equations on their own terms for the first time, mathematicians armed with computers are coming to new understandings of the way numbers can be used to represent the complex relationships between dimensions. Accepting that the surfaces in our world, from coastlines to clouds, exhibit the properties of both two- and three-dimensional objects (just what is the surface area of a cloud?) they came up with ways of working with and representing objects with fractional dimensionality. Using fractals and their equations, we can now represent and work with objects from the natural world that defied Cartesian analysis. We also become able to develop mathematical models that reflect many more properties of nature's own systems—such as self-similarity and remote high leverage points. Again, we find this renaissance characterized by the ability of an individual to reflect, or even affect, the grand narrative. To write the game.

Finally, our renaissance's answer to the printing press is the computer and its ability to network. Just as the printing press gave everyone access to readership, though, the computer and Internet give everyone access to authorship. The first Renaissance took us from the position of passive recipient to active interpreter. Our current renaissance brings us from a position of active interpretation to one of authorship. We are the creators.

Blog Cultures

So if we accept Rushkoff's premise of the present renaissance mimicking that of the previous Renaissance, how does our ability to create and to control our environments relate to the development of the blog? Many have argued that the growth of the blog is a revolution or at least major restructuring of mainstream media: that blogs are a powerful alternative news source and provide an effective check against the mass oligopolization of media organizations and concomitant reduction in content diversity.[4] But given the growth rates and decidedly personal nature of the majority of new blogs,[5] it's probably more significant that blogs represent the current renaissance's version of cross-cultural exchange. Safer and more efficient than circumnavigating the globe in either tall ships or spacecraft, blogs still allow for debate, exchange, and speculation between strangers, and the growth of new friendships, business deals, and partnerships that might never otherwise have been forged.

The rise of blogs as a medium for the creators of the modern communications renaissance has permitted an almost meteoric development of very specific cultural groups. Elsewhere in this edition are chapters focusing on niche blogging groups, but the culture of blogging itself has arisen as a key niche market. Celebrity bloggers are now a cult in their own right, with "A-list" identities such as Cory Doctorow, Meg Hourihan, Joi Ito, Jason Kottke, Glenn Reynolds, and Douglas Rushkoff (among many others) attracting major audiences simply because they are blogging. It helps, of course, to have a book or two under your belt to promote the image, but the act of blogging and the cult of blogging have produced their own unique market niche.

But is this blog culture impacting negatively upon the act of creativity? Is the growing blog culture a popularity contest among high-profile participants, where authorship by a few articulate pioneers is drowning out the voice of the people? Are blogging cultures in fact diluting the power and significance of the communications renaissance, and once again relegating most blog readers to active participants?

It was journalism in the nineteenth century that evolved as a Fourth Estate, providing stops and checks on the operations of the ruling classes and parliament. In her study on the changes taking place in news media today, Schultz noted that the public service principle of journalism made it the most important and powerful instrument of influence within a democracy.[6] The function of journalism and media reporting was designed to be informative, allowing the citizenry to respond to the issues of the day in a manner that ensured that the interests of the people were being considered. And the individual perspectives of the populace on all matter of issues, from religion to politics, began to have an impact on decision making. Thus multiplicity of individual perspectives of readers made possible by the printing press at the height of the Renaissance brought down the Church and spawned journalism simultaneously. There was a growing need to develop personalized responses to the world, and journalism could provide that voice for the masses.

But eventually the power of the media as a vehicle for the public voice and as a means of bringing down governing bodies was recognized, and commercial interests as well as political leaders sought to regulate the output of media institutions. Corporate mergers and acquisitions as well as formal and informal political affiliations emerged, and the number of media voices was reduced. Even the style of media reporting began to standardize. Globally, the perceived value of investigative journalism waned after the *Washington Post*'s association in the Watergate affair of the 1960s; by the turn of the millennium, in Western nations, representatives from the media were considered to be as trustworthy as used car salesmen. Even where the number of media channels in subscription television networks rose, the similarity of content production and centralization of news sources, as well as mass production of programming, was and is considered a reduction in media output quality.

Enter: blogs. Instead of relying on journalism as a trusted source of information and a means of checking the accuracy and authority of mainstream media, blogs provided a new shift: a change in perspective in news reporting and, more widely, in the creation of cultural content. The conflict that had emerged in mainstream media, of providing commercially viable content at low cost while also meeting an acceptable standard of news reporting, was deemed irrelevant in the blogosphere, because content was created at nearly zero cost for all authors, and the commentary system provided an immediate channel for response. Like the change in perspective experienced during the high Renaissance, the significance of blogs added a new dimension to our understanding of the world as previously experienced through print and electronic media channels.

But this supposed change in perspective is worth questioning. What we are seeing with the development of a blogging culture may not be as democratizing as the idealists would have us believe. The technology of blogging itself encourages centralization through networks of similar thinkers and debaters. David Weinberger has explored some of this territory in his work, describing the conversations erupting online through the vehicle of blogs and other social software tools as "small pieces, loosely joined,"[7] but it's worth exploring just how loosely these "pieces" of data—news, cultural content creation, business idea generation, legal mediation, learning facilitation, and so on—are joined.

Since the release of blogging software at the turn of the millennium, blogs have simply replaced older "home pages" of technology advocates and netizens alike. As the technology itself evolved, and articulate pioneer bloggers began to list "blogrolls" of other bloggers they read and advocated, and concepts like *Friendster* and *LiveJournal*'s Friends networks developed, there emerged—almost by accident—blog communities. Very quickly, a power law distribution of blog popularity emerged, and the significance of some blogs dramatically overshadowed the majority.[8] This did not mean that new blogs were prevented from rising to prominence—there have been several instances of blogs that have grown in influence over time—but the development of a hierarchy of blog significance did emerge, and syndicates of bloggers achieved a degree of influence approaching that of mainstream media. Gill cautioned that any attempt to measure the specific influence of blogs (based on links and audience numbers) is inherently flawed, because the competing methodologies for calculating these linkages and audiences are so complex. She also notes that other research questions about the relative influence of cross-linking through direct in-blog references as opposed to blogrolls, as well as the frequency of blog entries and first-mover advantage of early bloggers, will all impact on any degree of influence. But she also concludes that regardless of these measurement issues, there is no doubt that certain "A-list" blogs are so influential that they spawn other blog discussions and create a pyramid structure of influence.[9]

So while the functionality and power of negotiation inherent to blogs may well be providing us with a new perspective and a new means of understanding the world, there are even now the seeds of centralization of blog significance at work. Just as mainstream media have suffered oligopolization and centralization of control and voices, blogs may be considered to be oh-so-gradually moving down the same path of centralization.

Of course, the age of blogs competing with mainstream media as one of the most important instruments of influence in a democracy is a long way off, at least. Blogs may at present offer individuals and organizations a much more efficient manner of communicating, reporting, and negotiating content, but they are not yet a great threat to mainstream media, nor are they likely to bring down governments. Indeed, it could also be argued that blog centralization that is occurring could better be described as information fragmentation—even fracturing. As cults of "A-list" bloggers emerge with very specific agendas, and these blogs in turn spawn similar blogs (small pieces, loosely joined), what emerges is not a cyber-utopia of political debate and an agora for democratic practice, but a series of disconnected streams of information aggregation and discussion, highly specialized and politically specific. Rather than acting as a forum for dispassionate deliberation, pockets of political opinion can emerge, and participants of such blog cults positively reinforce their own ideals without consulting alternative arguments. In such an environment, blogs are no renaissance in communication, but merely an instrument of apartheid for individual perspectives.

The Real Threat of Blogging

> **» Douglas Rushkoff — The Real Threat of Blogs**   (5 Sep. 2004)[10]
>
> I believe that the most dangerous thing about blogs to the status quo is that so many of them exist for reasons other than to make money. A thriving community of people who are engaged for free, to me, have a certain authority that people doing things for money don't.
>
> Writing a book for money is always suspect. (Disclosure to all: I have written books for money and for free.) Writing it for free is very different—and might still be suspect, but for other reasons.
>
> What made the early Internet so very threatening to the mainstream media was not just the new opinions being expressed, but the fact that people were spending hours of their lives doing something that didn't involve production or consumption in the traditional market sense. Families with Internet connections were watching an average of nine hours less commercial programming each week.
>
> The threat of rave culture was that it was an alternative economy. The kids were no longer going to the mob-run nightclubs, the police weren't getting their cut, and the liquor distributors weren't making any money. Those of us involved in rave—or at least many of us—didn't realize that's why they were such a threat.
>
> Likewise, I believe the greatest power of the blog is not just its ability to distribute alternative information—a great power, indeed—but its power to demonstrate a mode of engagement that is not based on the profit principle.
>
> Admittedly, many people need to try to make money any way they can. And many people who insist on making their money by writing, but can't do so in the current commercial writing space, will attempt to do so on their blogs. I think that they will learn, as I have, what is so valuable about keeping certain areas of one's life and work

market-free if at all possible. Even if it means getting a day job, which many of us have done in order to support our work. I have been very lucky in my ability to craft my messages into forms that publishers will pay for. But the more integrity I get (and the more market-driven the book industry gets) the harder that is to do. Indeed, the book industry used to use criteria other than marketability in picking what to publish. Sometimes, editors would publish books that only broke even, because they happened to like them. Those editors are few and far between, now, because they don't make as much money for their companies, and the values have changed.

Conclusions

There is growing debate over business models for blogging, with the commercial sector keen to consider means of profiting from the blogging act. Of course, this business focus has spawned much concern among blogging "purists" about the effects of blog advertising on the validity of blogs as a source of critical debate. However, Rushkoff's discussion above is not focused on advertisements in blogs, but on the power of blogging as an activity which is largely disengaged from commercial media. Rushkoff's notion of the real threat of blogs goes some way toward alleviating the potential problems that could arise from a growing centralization (or fragmentation) of voices. If the real "threat" of blogging is its challenge to mainstream media as a nonprofit activity, then a very old problem arises: how do businesses access consumers in an age of content generation by the people and for the people? Any fragmentation of voices occurring in the blogosphere will be secondary to the primary concern of sales and marketing for commercial players. If blogs threaten audience sizes and types for mainstream media, then the pricing of advertising for media will decline, and costs of production for advertising-supported media will consequently rise. Businesses will seek out new ways of accessing audiences either through *Google*-esque search engine and context-specific Website advertising, or through some other manner of contextual information aggregation.

This threat is real and palpable for businesses today. Research is being conducted on new models for television and interactive media advertising across the globe precisely because businesses still need a means of providing information about products and services to potential consumers. And as blogs inherently provide content that audiences find engaging and interesting, businesses are also considering how they can use blogs as a vehicle for accessing those highly specialized audiences.

Chris Anderson has explored this in his ideas about "the Long Tail": an entertainment or content industry that is characterized by millions of niche markets at the "shallow end of the bitstream."[11] Blogs fit perfectly into this theory of business development, because they respond so dynamically to the changing wants and needs of what Bruns has called "produsers."[12] Being loosely collected communities of content creators and readers, blogs provide for businesses an opportunity to access niche interest groups while also collecting data on the changing interests of those groups. Anderson argues that the "long tail" actually facilitates development of diverse content, reversing the trend of content homogenization in mainstream media. Business generated in the "long tail" doesn't seek to replace mass-market fare, but augments it for niche interests. Similarly, blogs don't replace mass media content, but they augment consumer content production. Blogs have what Tapscott calls "digital capital":[13] the combination of employee (blogger) and customer (blog reader) knowledge, customer relationship capital (blog readership understanding and growth), and knowledge maximization for business efficiencies (use of that understanding to grow blog influence and readership). Bloggers develop an understanding of their readership through commentary systems and can dynamically affect their readership and influence through cross-linking posts and contributing to other blogs. Thus, blogs are uniquely suited to business development in the "long tail," because blogs are content vehicles for niche interests, and bloggers have the digitial capital that businesses seek.

The trouble with this emergence of blogging as a facilitator for business is that what Rushkoff describes as the "real threat" of blogs begins to weaken as blogs formally or informally adopt business models for connecting businesses with potential consumers. And with highly specialized business markets erupting in the "long tail," the threat of blogs as non-profit ventures declines, and market segmentation—and fragmentation of perspectives—once again rises.

We're left wondering what kind of communication renaissance has really emerged, and who are the greatest beneficiaries of the blogging revolution.

NOTES

1. Tom Coates, "Second Sight," *The Guardian*, 28 Aug. 2003.

2. Mark Poster, "CyberDemocracy: Internet and the Public Sphere," Lecture at the University of California, Irvine, 1995, http://www.uoc.edu/in3/hermeneia/sala_de_lectura/mark_poster_cyberdemocracy.htm (accessed 16 Nov. 2005).

3. Excerpted from Douglas Rushkoff, "Renaissance *Now?*" *Douglas Rushkoff*, 13 June 2002, http://www.rushkoff.com/2002/06/renaissance-now.php (accessed 16 Nov. 2005). This article also touches on issues covered in more detail in Douglas Rushkoff, *Open Source Democracy: How Online Communication Is Changing Offline Politics* (London: Demos, 2003), http://www.demos.co.uk/opensourcedemocracy_pdf_media_public.aspx (accessed 22 April 2004).

4. See, among others, the chapter by Axel Bruns in this text, and David Abrahamson, "From the Many, to the Many: The Journalistic Promise of Blogs," *Journal of Magazine and New Media Research* (Summer 2005).

5. See, in particular, David Sifry, "State of the Blogosphere, August 2005, Part 5: The A-List and the Long Tail," *David Sifry's Alerts*, Aug. 2005, http://www.technorati.com/weblog/2005/08/39.html, 2005 (accessed 16 Nov. 2005).

6. Julianne Schultz, *Reviving the Fourth Estate: Democracy, Accountability and the Media* (London: Cambridge UP, 1998).

7. David Weinberger, *Small Pieces, Loosely Joined* (London: Cambridge UP, 1998).

8. Clay Shirky, "Power Laws, Weblogs, and Inequality," *Clay Shirky's Writings about the Internet: Economics & Culture, Media & Community, Open Source*, 2 Oct. 2003, http://www.shirky.com/writings/powerlaw_weblog.html (accessed 20 Feb. 2004).

9. Kathy Gill, "How Can We Measure the Influence of the Blogosphere?" *WWW2004 Conference Proceedings*, University of Washington, 2004.

10. Excerpted from Douglas Rushkoff, "The Real Threat of Blogs," *Douglas Rushkoff*, 5 Sep. 2004, http://www.rushkoff.com/2004/09/real-threat-of-blogs.php (accessed 16 Nov. 2005).
    Issues addressed both in this article and in the previous excerpt are also explored in more detail in Douglas Rushkoff, *Get Back in the Box: Innovation from the Inside Out* (London: CollinsBusiness, 2005).

11. Chris Anderson, "The Long Tail," *Wired Magazine* 12.10 (Oct. 2004).

12. See Axel Bruns, "Some Exploratory Notes on Produsers and Produsage", *Snurblog*, 3 Nov. 2005, http://snurb.info/index.php?q=node/329 (accessed 4 Nov. 2005).

13. Don Tapscott, *Digital Capital: Harnessing the Power of Business Webs* (Cambridge, MA: Harvard Business School P, 2000).

CHAPTER TWENTY-TWO

What's Next for Blogging?

Axel Bruns

This book emerged out of a realization that amid a glut of "blogging for dummies"-style guides, there are still far too few in-depth studies of how blogs and blogging are being adopted across a wide range of fields. With this volume, we've done our best to begin this process of charting the uses of blogs right across the spectrum, from blogging for personal expression to blogging for commercial gain, and beyond. Even as we put the final touches on this collection, however, we're well aware that blogging is a highly movable feast: someone, somewhere, in the vast and complex network of the blogosphere is currently developing the Next Big Thing in blogging just as you read this.

Who might these people be, and what are they working on? How will their inventions filter through the network and be adopted across the world of blogs? What will determine their eventual success or failure? Indeed, who drives the continuing and continuous development of the blogosphere as a social, intellectual, creative, and perhaps in part even commercially viable network? A study of the individual uses of blogs—their history, current status, and future trajectory—provides some possible answers to these questions; but to understand better what the blogosphere itself may think, we approached a few genuine "A-list" bloggers to give us their points of view.[1]

Beyond "Blogging"

By now there seems to be a fairly widespread sense of agreement that the time of describing "blogging" as a unified practice or genre of textual production is rapidly passing. Political blog commentator Tim Blair notes that "people speak of personal blogs, political blogs, pet blogs, satire blogs," and many other forms of blogging, and feminist academic blogger Clancy Ratliff agrees:

even now "blogging" is only slightly more meaningful a term than "publishing." With blogging, the special meaning is that what you're doing is self-publishing without an editor or other gatekeeper. When I first started studying weblogs in 2002, I quickly realized that making claims or generalizations about weblogs was like trying to say that X is true of all books: Maybe you can make a claim about all books, but it wouldn't be a very meaningful one.

Indeed, recent attempts at summarizing blogs and blogging have returned to focus on the few remaining unifying characteristics of blogs: Jill Walker's definition of "blog" for the *Routledge Encyclopedia of Narrative Theory*, for example (probably one of the first narrative encyclopedias to recognize blogs as an important new phenomenon), begins by stating that

> a weblog, or *blog, is a frequently updated website consisting of dated entries arranged in reverse chronological order so the most recent post appears first (see temporal ordering). Typically, weblogs are published by individuals and their style is personal and informal. Weblogs first appeared in the mid-1990s, becoming popular as simple and free publishing tools became available towards the turn of the century. Since anybody with a net connection can publish their own weblog, there is great variety in the quality, content, and ambition of weblogs, and a weblog may have anywhere from a handful to tens of thousands of daily readers.[2]

What is noticeable about this description is that it focuses in good part on technological features of blogs, rather than attempting to define generic markers—this may now be the last available option if a universally applicable description of the essence of blogs is required. Indeed, Clancy Ratliff notes that some commentators have suggested "that a weblog is not a genre; it's a technology. I don't know if I agree that generic conventions are all that separate from specific writing technologies." Clearly the technological features of publishing technologies also help determine what genres may be possible within their confines; but at the same time technologies are also shaped by the social needs that are present in contemporary culture and may drive the rise and fall of particular genres of expression.

It appears obvious at this point that a key attraction to the blogosphere is, and remains, the potential for individual and informal expression and ungatekept self-publishing which both Ratliff and Walker identify, and it seems safe to assume that in any future developments of blogging genres, this form of user-led content production, or produsage, will continue to play a significant role. The ability to link to and comment on content found on other blogs or elsewhere on the Web also remains a crucial aspect, and tools both for identifying interesting links as well as for analyzing the linkage patterns of bloggers more generally continue to multiply, even if some mechanisms for remote commenting and annotation of blogs and Websites—such as trackbacks—have been severely compromised because of their misuse by spammers.

Technological and Generic Futures for Blogging

From a technological point of view, however, we may soon see an increase in the forms such publishing and commenting might take. As it has happened in the history of other communications forms (from telegraph to telephone, from print newspaper to glossy magazine, from HTML1.0 Web to the multimedia Web as we know it today), it is likely that blogging will continue to embrace a wider range of media forms. As Ratliff notes, "with the rise of podcasting, vlogging, large collaborative audio projects like [murmur], the joining together of interactive online art and graffiti (Grafedia.net), and who knows what else in the pipeline, there are going to be plenty of other ways to publish besides keeping a text-based weblog."

Whether such technologies spawn entirely new blogging genres remains to be seen, however—and this may also crucially depend on the emergence of key new tools, much in the same way that the availability of *Blogger* and *LiveJournal* significantly contributed to the growth of the blogosphere. Vlogging, or video blogging, for example, has been technologically possible for some time now, but has yet to be widely adopted (new sites such as the "Audio, Video, and Podcast Publishing Service" *Audioblog.com* might now begin to further this process), and even photoblogging remains relatively underdeveloped even in spite of the availability of relatively cheap camera phones and sites such as *Flickr*.

But even the availability of technological tools and supports does not guarantee the eventual success of a blogging genre—many bloggers might continue to find it easier to draft a pithy, funny, or insightful statement on their field of interest for a text-based blog than to deliver the same as an impromptu rant into a Webcam or microphone video or audio blog. Blog readers, too, may well prefer texts that can be read quickly ("scanning" time-based multimedia in the same way that texts can be speed-read remains difficult) and silently (audio and video blogs may attract unintended listeners in public, in the office, or at home).

Perhaps the greater impetus for the continuing development of blogs in all their forms is driven by the evolution of genres rather than technologies, then. As our contributors in this book have shown, blogs are being adopted for (and, frequently, adapted to) an increasingly wide variety of uses, and it is likely that such developments will continue to cross-influence one another in often unexpected ways. The access controls for specific forms of corporate "dark blogs," for example, are perhaps not all that dissimilar from the tools for restricting reader access to specific in-groups as personal bloggers in *LiveJournal* might use them, and the variations to blogging genres that are being trialed in sites such as *Cyworld* could well instruct wider changes in the blogosphere.

The point here is not to speculate on where such generic developments may lead us, and how they might affect the nature of the blogosphere as a whole, but simply to use these observations as a motivation both to continue a close study of what developments do occur in various genres of blogging, and (as researchers as well as bloggers) to keep experimenting with the uses of blogs. It is likely that we will find unexpected successes as well as failures in using blogs, and these, too, will be instructive for further development. Widespread deployment of blogs and related genres and technologies may indeed lead to effects well beyond the blogosphere itself, and as Clancy Ratliff suggests, "the term 'citizen media' is inclusive enough to encompass all of these technologies while still retaining the self-publishing, unedited by larger organizations aspect of this phenomenon."

Mainstreaming Blog Innovations

Terms such as "citizen media" can obscure the question of what drives the deployment of such new technologies and uses, however. Do new approaches to blogging simply spread through the blogosphere by diffusion, like liquid being sucked up into a sponge, or is the process driven by individual adopters (and if so, who are they)? Are there even external influences at work here? Ratliff notes, for example, that public perceptions of the blogosphere are hardly representative:

> I ... don't think one can talk about the blogosphere without talking about its uptake in mainstream media and its representations in popular culture. In the United States, I find this to be a problem. The bloggers who get the most positive attention from major news organizations as well as opportunities to publish in other venues tend to be white men. Political, filter-style weblogs are masculinized, personal, diary-style weblogs are feminized, and the two types are overly bifurcated. Personal weblogs ... are represented as narcissistic and confessional, and blogging has also been portrayed as activity associated with stereotypical teenaged girls.

In this environment, would-be bloggers setting up their own blogs and developing their own style of writing and publishing blog entries may therefore feel obliged to try overly conventional, even stereotypical, approaches to blogging, rather than finding their own voice and exploring one of the many *other* genres of blogging that are less prominent in the mainstream media, while some writers seeking a medium within which to express their thoughts might not regard blogging as an opportunity because "their" genre of blogging is hardly visible.

Similarly, if mainstream media portrayals of blogs suggest a false uniformity for blogging styles, then so does an exaggerated focus on "A-list" bloggers. Ratliff notes that "the bloggers who get the most traffic are in a privileged position. They amplify the voices of the writers to whom they link and expose their writing to a much larger audience." And in the same way, by predominantly linking to blogs that operate in the same genres as they do, they might also create a sense of conformity with regard to what a "proper" blog should look like. Much as Ratliff feels that "it's in the interest of a democratic blogosphere to seek out and bring in minority positions and issues," then, it may also be in the interest of a diverse and developing blogosphere that "A-list" bloggers highlight new approaches to blogging and themselves experiment with new uses of blogs.

However, in spite of the significant interest in "A-list" bloggers (both within the blogosphere itself, and in the wider mediasphere), it is also important to recognize that "A-list" status is fleeting rather than fixed. To some extent, even "A-list" bloggers are only ever as good as their next blog posting; "if a blogger publishes something he or she knows to be inaccurate, or is proven later to be inaccurate, it can take a long time to live that down. Writers who take blogging seriously do not want to lose their credibility or their audience, and they tend to correct publicly any errors they make," Ratliff notes. Further, as Clay Shirky has demonstrated, there exists an exponential or "power law" distribution of readership ranging from a very small number of very well-known "A-list" bloggers to a vast majority of very rarely accessed blogs[3]—but at the same time, it would be a mistake to simply dismiss the collective importance of bloggers outside of the "A-list"; as Ratliff argues, "it's not as though if you're not on 'the A-list,' whatever that means ... , no one will read your weblog, or that if you start a weblog, it will be all that much harder to garner an audience than it was four or five years ago."

Instead, in keeping with the history of blogging as a user-led, grassroots, self-published medium, success, both for bloggers and for blog technologies and uses, is determined not solely by whether they are recognized by those keeping the gates of existing publications (be they "A-list" bloggers or other media proprietors), or by those controlling the purse-strings of technology companies, but importantly also by whether individual bloggers can make themselves heard and demonstrate the validity of their approach. Thus, ultimately, Clancy Ratliff suggests, "if one thinks of 'democratic' in this sense as having equal opportunity to speak and be heard, I believe the blogosphere can be democratic, at least to an extent, if the blogger reaches out and joins a conversation, writes herself into the network ... by linking to other weblogs, commenting at other weblogs, and making use of trackback." If such attempts at spreading the word and establishing new approaches for blogging are successful, it is likely that they will become recognized and adopted throughout the blogosphere.

This, then, is a further indication of the strange new status of the blogosphere as a distributed environment for user-led innovation, or produsage.[4] Or, as author and blogger Neil Gaiman puts it, "the blogosphere is not organised, but it's really well disorganised."

NOTES

1. Email interviews with Clancy Ratliff, Tim Blair, Neil Gaiman, and other bloggers not quoted here were conducted throughout 2005 by research assistant Ian Rogers. Clancy Ratliff also blogged her responses at *CultureCat: Rhetoric and Feminism*: "Interview for *Uses of Blogs*," 24 July 2005, http://culturecat.net/node/876 (accessed 7 Nov. 2005). Our sincere thanks to all the bloggers who responded to our questionnaire.

2. Jill Walker, "Final Version of Weblog Definition," *jill/txt*, 28 June 2003, http://huminf.uib.no/~jill/archives/blog_theorising/final_version_of_weblog_definition.html (accessed 7 Nov. 2005).

3. Clay Shirky, "Power Laws, Weblogs, and Inequality," *Clay Shirky's Writings about the Internet: Economics & Culture, Media & Community, Open Source*, 2 Oct. 2003, http://www.shirky.com/writings/powerlaw_weblog.html (accessed 20 Feb. 2004).

4. See Axel Bruns, "Some Exploratory Notes on Produsers and Produsage", *Snurblog*, 3 Nov. 2005, http://snurb.info/index.php?q=node/329 (accessed 4 Nov. 2005).

Contributors

Axel Bruns (a.bruns@qut.edu.au) teaches and conducts research in the Creative Industries Faculty at Queensland University of Technology in Brisbane, Australia. He is co-founder and General Editor of M/C–Media and Culture (www.media-culture.org.au). Axel has published extensively on new models for content production and editing in online journalism, and is the author of *Gatewatching: Collaborative Online News Production* (New York: Peter Lang, 2005), the first comprehensive study of the latest wave of online news publications that employ open source ideals in their use of collaborative production tools.

Axel Bruns

He is currently involved in a number of research projects that extend this work to a wider study of produsage—distributed, collaborative, and user-led content production and innovation—in a variety of contexts, also focusing on the question of how to embrace the principles of produsage in the development of engaging learning environments. More information on his research, as well as selected writings, can be found on his Website at snurb.info.

Joanne Jacobs (joanne@joannejacobs.net) is a Project Manager at the Australasian Cooperative Research Centre for Interaction Design (www.interactiondesign.com.au) where she oversees research on collaborative technologies and innovation processes. She is also a private information and communications technology consultant and a Lecturer in the MBA program at the Brisbane Graduate School of Business at Queensland University of Technology.

Joanne Jacobs

Her areas of research expertise are Internet regulation, digital television, social software, electronic publishing, e-marketing, and telecommunications policy. In all these areas, her work has focused on participation, diversity, and access to the technologies of communication. She has written widely on these areas, and has drawn on her consultancy work to develop practitioner-oriented reviews and guidelines. She has developed policies for both the public and private sector relating to digital delivery and virtual participation, and she has acted in executive roles in a number of organizations in the information and communication industry sector. She is also an advisory member to the National Forum, Inc., publisher of the highly successful *On Line Opinion* (www.onlineopinion.com.au), a Web-based journal of social and political debate. Her blog can be found at joannejacobs.net.

Mark Bahnisch Mark Bahnisch (mbahnisch@gmail.com) is a Lecturer in Sociology in the School of Arts, Media and Culture at Griffith University. He has published on political and social theory, Australian and international politics, the sociology of deviance, industrial relations, organizational sociology and sociology of religion. Mark also works as a consultant in the areas of employment relations, policy and social analysis and has consulted to the Australian and Queensland governments and public and private sector organizations. His blog is located at larvatusprodeo.redrag.net.

Jean Burgess Jean Burgess (je.burgess@qut.edu.au) is a doctoral candidate in the Creative Industries Faculty at Queensland University of Technology, Australia. Her doctoral research project "Vernacular Creativity and New Media" investigates the implications of amateur content creation for the politics of cultural participation, focusing particularly on the digital storytelling movement. She regularly lectures and publishes on cultural studies, popular music, and new media. Her Weblog is located at hypertext.rmit.edu.au/~burgess.

Suw Charman Suw Charman (suw.charman@gmail.com) is a renowned social software expert who specializes in the application of blogs and wikis in business. An independent consultant, she works with companies in the United Kingdom and the United States, advising on the use of blogs both behind the firewall and as marketing and external communications tools. Suw frequently speaks at conferences and seminars and is currently writing a book on dark blogs.

A well-known blogger, Suw keeps a number of blogs, including *Chocolate and Vodka* (chocnvodka.blogware.com), *Strange Attractor* (corante.com/strange), and *Blogiculum Vitae* (suw.org.uk). She is also Executive Director of the Open Rights Group (openrightsgroup.org), a British digital rights organization.

Jaz Hee-jeong Choi Jaz Hee-jeong Choi (jaz@nicemustard.com) is a Ph.D. candidate in the Creative Industries Faculty at Queensland University of Technology. Her research interests are in digital communication, particularly the ways in which various forms of digital communication are developed, established, and utilized in an Asian context. Her current research is a triangulation study of Japan, South Korea, and the People's Republic of China. The study explores how popular culture is communicated across these cultures in the digital realm, and investigates the circumstantiality and potentiality of this rapidly evolving cultural network based on common techno-cultural and communicative denominators. Her Website is located at www.nicemustard.com.

Trevor Cook (trevor.cook@gmail.com) is a director with Jackson Wells Morris, a Sydney PR firm,
Trevor Cook
where he has worked for the past decade. Trevor is an honors graduate in economics and spent fifteen years working in national government as a researcher, policy adviser, and program director.

Trevor has been blogging since November 2003, and his main blog is *Corporate Engagement* (trevorcook.typepad.com). In 2004 he initiated Global PR Blog Week (www.globalprblogweek.com), now an annual event.

James Farmer (james@incsub.org) is an education designer and social software consultant living and
James Farmer
working in Melbourne, Australia. He co-organized the first blogging conference in the southern hemisphere, "Blogtalk Downunder" (incsub.org/blogtalk), runs the blog consultancy Blogsavvy (blogsavvy.net), and is the founder of the blogging services *edublogs.org*, *learnerblogs.org*, and *uniblogs.org*.

In his spare time he researches and lectures in education design at Deakin University and occasionally manages to fit in a walk by the river and some very amateur DIY. He finds it a bit strange writing about himself in the third person, but does it anyway. His blog is located at incsub.org.

Brian Fitzgerald (bf.fitzgerald@qut.edu.au) is a well-known intellectual property and information
Brian Fitzgerald
technology lawyer and Professor and Head of the Law School at Queensland University of Technology. He is co-editor of one of Australia's leading texts on e-commerce, software and the Internet—*Going Digital 2000*—and has published articles on Law and the Internet in Australia, the United States, Europe, Nepal, India, Canada, and Japan. His latest (co-authored) books are *Cyberlaw: Cases and Materials on the Internet, Digital Intellectual Property and E Commerce* (2002); *Jurisdiction and the Internet* (2004); and *Intellectual Property in Principle* (2004). Over the past four years Brian has delivered seminars on information technology and intellectual property law in Australia, Canada, New Zealand, the United States, Nepal, India, Japan, Malaysia, Singapore, Norway, and the Netherlands.

Brian is also a Chief Investigator in the newly awarded ARC Centre of Excellence on Creative Industries and Innovation at QUT. His current projects include work on digital copyright issues across the areas of Open Content Licensing and the Creative Commons, Free and Open Source Software, Fan Based Production of Computer Games, Licensing of Digital Entertainment, and Anti-Circumvention Law. Brian is a Project Leader for Creative Commons in Australia. From 1998 to 2002 Brian was Head of the School of Law and Justice at Southern Cross University in New South Wales, Australia, and in January 2002 was appointed as Head of the School of Law at QUT in Brisbane, Australia. His Website is located at www.law.qut.edu.au/about/staff/lsstaff/fitzgerald.jsp.

Gerard Goggin   Gerard Goggin (g.goggin@uq.edu.au) is an ARC Australian Research Fellow at the Centre for Critical and Cultural Studies, University of Queensland, who has published widely on Internet, mobiles, and disability. He is the author of *Cell Phone Culture: Mobile Technology in Everyday Life* (Routledge, 2006) and *Digital Disability: The Social Construction of Disability in New Media* (with Christopher Newell; Rowman & Littlefield, 2003), as well as editor of *Virtual Nation: The Internet in Australia* (University of NSW Press, 2004). His Website is located at gerardgoggin.net.

Melissa Gregg   Melissa Gregg (m.gregg@uq.edu.au) is a Postdoctoral Research Fellow in the Centre for Critical and Cultural Studies at the University of Queensland, where she also lectures in the School of English, Media Studies, and Art History.

She is the author of *Investing in Cultural Studies: Politics, Affect & the Academy* (Palgrave, 2006) and has published in a variety of academic journals including *Cultural Studies*, *Cultural Studies Review*, *Continuum*, and *Antithesis*. Melissa's blog *Home Cooked Theory* (hypertext.rmit.edu.au/~gregg) reflects her research interests, which include intellectual history, contemporary politics, and workplace culture. Further details are also available at the CCCS Website (cccs.uq.edu.au/index.html?page=16194&pid=16136).

Alexander Halavais   Alexander Halavais (alex@halavais.net) is an Assistant Professor of Informatics and Communication at the University at Buffalo (State University of New York). He directs the Masters in Informatics program, which addresses the social and organizational aspects of information and communication technologies. His research looks at "social computing" and its impact on social change, journalism, education, and public policy. He has recently edited a reader, *Cyberporn & Society*, and teaches a popular undergraduate course on the same topic. His blog may be found at alex.halavais.net.

Paul Hodkinson (p.hodkinson@surrey.ac.uk) is Lecturer in Sociology at the University of Surrey, **Paul Hodkinson**
where he teaches in the Sociology, Culture, and Media program. His research has consistently been concerned with the relationship between media and patterns of individual and collective identity among young people. Such issues are explored via a comprehensive reworking of the notion of subculture in his book *Goth: Identity, Style and Subculture* (Berg, 2002), a publication that gave rise to numerous national media reviews and interviews. He has also published journal papers and book chapters focused on the implications of Internet communication for the form taken by contemporary youth cultures. Paul is one of the co-convenors of the British Sociological Association Youth Study Group and recently co-organized the "Young People and New Technologies" conference at the University of Northampton in the United Kingdom. More information about Paul and his work is available at www.paulhodkinson.co.uk.

Adrian Miles (adrian.miles@rmit.edu.au) is a Lecturer in Cinema and New Media at RMIT Uni- **Adrian Miles**
versity, Melbourne, and was formerly Senior Researcher in New Media at the InterMedia Lab, University of Bergen. He is an internationally recognized theorist and creator of hypermedia, and his academic work has been published in numerous peer-reviewed publications. He also is a networked interactive video developer, and his applied projects have been exhibited internationally.

Adrian's research practice concentrates on hypermedia, interactive narrative, and the development of media annotation tools to facilitate audiovisual research in humanities contexts. He has a specific theoretical interest in link poetics and conceptualizing an aesthetics of interactive video. This work utilizes the cinema theory of Gilles Deleuze and seeks to apply it in novel ways to rethink notions of interactivity, the user, and the text. His creative practice explores the aesthetics and affordances of networked, distributed video and the possibilities for new forms of distribution and expression that this may enable. His blog is located at hypertext.rmit.edu.au/vlog.

Tim Noonan   Tim Noonan (tnoonan@softspeak.com.au) holds a B.A. in cognitive psychology and special education. He has specialized in the disabilities/technology field for over twenty years, with a strong focus on accessible and useable design. Since 1995 Tim has consulted widely for industry, government, and NGOs through SoftSpeak Consulting. He is a frequent guest on radio and television, where he engagingly examines issues of social inclusion and access to information and emerging technologies. His Website is located at www.timnoonan.com.au.

Damien O'Brien   Damien O'Brien (d2.obrien@qut.edu.au) holds a Bachelor of Laws and is a research assistant in the Faculty of Law at Queensland University of Technology, Brisbane.

Ian Oi   Ian Oi (ian.oi@corrs.com.au) is a partner at Corrs Chambers Westgarth lawyers in Canberra, Australia. His practice focuses on intellectual property, information technology and telecommunications, and strategic procurement. Formerly a senior attorney and negotiator with IBM in Australia and a special counsel with another large Australian firm, Ian has extensive experience advising on these areas for clients in the public and private sectors. He has acted for suppliers and customers in some of Australia's most significant technology-related projects in recent times, and has particular expertise in dealing with large outsourcing transactions, complex software development, IP commercialization, and open source software and open content development, management, and distribution.

Ian is regularly invited to speak around Australia on issues in the above areas. His recent presentations have been on topics such as government procurement after the Australia-U.S. Free Trade Agreement, recent developments in online contracting, open source software projects in the Australian public sector, the legal implications for electronic archival initiatives of "digital amnesia," and Creative Commons initiatives in Australia. More information about his work can be found at www.corrs.com.au/corrs/website/web.nsf/Content/OiIan.

John Quiggin   John Quiggin (j.quiggin@uq.edu.au) is a Federation Fellow and Professor in Economics and Political Science at the University of Queensland. He is prominent both as a research economist and as a commentator on Australian economic policy. He has published over 750 research articles, books, and reports in fields including environmental economics, risk analysis, production economics, and the theory of economic growth. He has also written on policy topics including unemployment policy, micro-economic reform, privatization, competitive tendering, and the management of the Murray-Darling river system.

John has been an active contributor to Australian public debate in a wide range of media. He is a regular columnist for the *Australian Financial Review*, to which he also contributes review and feature articles. He frequently comments on policy issues for radio and television. He was one of the first Australian academics to present publications on a Website (now at www.uq.edu.au/economics/johnquiggin). In 2002, he commenced publication of a Weblog (now at johnquiggin.com) providing daily comments on a wide range of topics.

Douglas Rushkoff (rushkoff@well.com) is the winner of the first Neil Postman Award for Career Achievement in Public Intellectual Activity. He is an author, teacher, and documentarian who focuses on the ways people, cultures, and institutions create, share, and influence each other's values. He sees "media" as the landscape where this interaction takes place, and "literacy" as the ability to participate consciously in it. His ten best-selling books on new media and popular culture have been translated into over thirty languages. They include *Cyberia*, *Media Virus*, *Playing the Future*, *Nothing Sacred: The Truth about Judaism*, and *Coercion*, winner of the Marshall McLuhan Award for best media book. Rushkoff also wrote the acclaimed novels *Ecstasy Club* and *Exit Strategy*, and the graphic novel *Club Zero-G*. He has just finished a book for HarperBusiness, applying renaissance principles to today's complex economic landscape, *Get Back in the Box: Innovation from the Inside Out*.

Douglas Rushkoff

Rushkoff founded the Narrative Lab at NYU's Interactive Telecommunications Program, and lectures on media, art, society, and change at conferences and universities around the world. He is Advisor to the United Nations Commission on World Culture, on the Board of Directors of the Media Ecology Association, The Center for Cognitive Liberty and Ethics, and a founding member of Technorealism. He has been awarded Senior Fellowships by the Markle Foundation and the Center for Global Communications Fellow of the International University of Japan. His Website and Weblog can be found at rushkoff.com.

Jane B. Singer (jane-singer@uiowa.edu) is an Associate Professor in the School of Journalism and Mass Communication at the University of Iowa in the United States. Her

Jane B. Singer

teaching and research interests are primarily in the area of online journalism, particularly the intersection of print and interactive media. She is especially interested in the sociology of online news work, or the people and processes behind the creation of online news, as well as in journalism ethics and political news coverage. Her research has appeared in a variety of publications, including *Journalism & Mass Communication Quarterly*, *Journalism Studies*, and *Journalism: Theory, Practice and Criticism*. She is currently a contributing editor to *Media Ethics* magazine and an editorial board member of *Journalism Studies* and the *Journal of Mass Media Ethics*.

Jane's professional experience includes ten years in the editorial department of what evolved into the Prodigy service. She was Prodigy's first news manager, in charge of one of the first around-the-clock news products ever to be delivered to people's homes through a computer. She also has five years' experience as a reporter and editor at three East Coast U.S. newspapers. Singer holds a Ph.D. in journalism from the University of Missouri-Columbia, an M.A. in liberal studies from New York University, and a bachelor's degree in journalism from the University of Georgia. Her Website is located at myweb.uiowa.edu/jsinger/.

Angela Thomas Angela Thomas (a.thomas@usyd.edu.au) is a Lecturer in English Education at the University of Sydney. Her research interests include young people's digital worlds, online role-playing, virtual identity, digital literature and literacies, and feminism and media studies. Recent publications include *e-Selves: Cyberkids, Literacy and Identity* (2006); *Digital Literacies* (2006, with Colin Lankshear and Michele Knobel); and *Children's Literature and Computer Based Teaching* (2005, with Len Unsworth, Alyson Simpson, and Jennifer Asha). Her blog is located at anya.blogsome.com.

Jill Walker Jill Walker (jill.walker@uib.no) is an Associate Professor in the Department of Humanistic Informatics at the University of Bergen in Norway, where she teaches digital media aesthetics and Web textuality. Jill's research areas include distributed narrative, electronic literature, networked art, and Weblogs. Her research blog is *jill/txt*, at jilltxt.net.

Bibliography

In developing the ideas presented in this book, the following key sources were useful to our contributors. Additional resources of specific relevance to individual forms of blogging are listed in the endnotes for each chapter.

Abrahamson, David, "From the Many, to the Many: The Journalistic Promise of Blogs," *Journal of Magazine and New Media Research* (Summer 2005).

Adar, Eytan, Li Zhang, Lada A. Adamic, and Rajan M. Lukose, "Implicit Structure and the Dynamics of Blogspace," Workshop on the Weblogging Ecosystem, 13th International World Wide Web Conference, May 18, 2004.

Alleyne, Mark D., *News Revolution: Political and Economic Decisions about Global Information* (Houndmills, UK: Macmillan, 1997).

Bausch, Paul, Matthew Haughey, and Meg Hourihan, *We Blog: Publishing Online with Weblogs* (Indianapolis: Wiley, 2002).

Berners-Lee, Tim, *Weaving the Web* (London: Orion Business Books, 1999).

Blood, Rebecca, *The Weblog Handbook: Practical Advice on Creating and Maintaining Your Blog* (New York: Perseus, 2002).

Bloom, David, "The Blogosphere: How a Once-Humble Medium Came to Drive Elite Media Discourse and Influence Public Policy and Elections," 2003, http://darkwing.uoregon.edu/~jbloom/APSA03.pdf (accessed 21 Sep. 2005).

Bolter, Jay David, *Writing Space: The Computer, Hypertext, and the History of Writing* (Hillsdale, NJ: Lawrence Erlbaum, 1991).

Bowman, Shane, and Chris Willis, *We Media: How Audiences Are Shaping the Future of News and Information*, http://www.hypergene.net/wemedia/weblog.php?id=P41 (accessed 25 Sep. 2005).

Brabazon, Tara, *Digital Hemlock: Internet Education and the Poisoning of Teaching* (Sydney: U of New South Wales P, 2003).

Bruns, Axel, *Gatewatching: Collaborative Online News Production* (New York: Peter Lang, 2005).

——, "Some Exploratory Notes on Produsers and Produsage," *Snurblog*, 3 Nov. 2005, http://snurb.info/index.php?q=node/329 (accessed 4 Nov. 2005).

Bucy, Erik P., and John E. Newhagen, eds., *Media Access: Social and Psychological Dimensions of New Technology Use* (Mahwah, NJ: Lawrence Erlbaum, 2003).

Bush, Vannevar, "As We May Think," *The Atlantic* 176.1 (1945).

Charman, Suw, "Dark Blogs: Case Study 01: A European Pharmaceutical Group," 13 June 2005, http://www.suw.org.uk/files/Dark_Blogs_01_European_Pharma_Group.pdf (accessed 3 Nov. 2005).

Coates, Tom, "Weblogs and the Mass Amateurization of (Nearly) Everything," *Plasticbag.org*, Sep. 2003, http://www.plasticbag.org/archives/2003/09/weblogs_and_the_mass_amateurisation_of_nearly_everything.shtml (accessed 17 July 2005).

Cohen, Kris, "A Welcome for Blogs," *Continuum Special Issue: Counter-Heroics and Counter-Professionalism in Cultural Studies* 20.2 (June 2006).

Doctorow, Cory, "My Blog, My Outboard Brain," *O'Reilly DevCenter Articles* (2002), http://www.oreillynet.com/pub/a/javascript/2002/01/01/cory.html (accessed 27 Oct. 2005).

Downes, Stephen, "Educational Blogging," *EDUCAUSE Review* 39.5 (2004), pp. 14–26.

Ebanks, Ken, "Into the Blogosphere: Managing the Risks and Rewards of Employee Blogging," *The Computer and Internet Lawyer* 22.1 (Oct. 2005).

Eco, Umberto, *Travels in Hyperreality* (New York: Harvest Books, 1986).

Efimova, Lilia, and Aldo de Moor, "Beyond Personal Webpublishing: An Exploratory Study of Conversational Blogging Practices," *Proceedings of the 38th Hawaii International Conference on System Sciences*, 2005.

——, Sebastian Fiedler, Carla Verwijs, and Andy Boyd, "Legitimized Theft: Distributed Apprenticeship in Weblog Networks," *Proceedings of I-KNOW04*, Graz, Austria, 2004.

——, and Stephanie Hendrick, "In Search for a Virtual Settlement: An Exploration of Weblog Community Boundaries," 2005, https://doc.telin.nl/dscgi/ds.py/Get/File-46041 (accessed 3 Nov. 2005).

Electronic Frontier Foundation, "Legal Guide for Bloggers," 6 July 2005, http://www.eff.org/bloggers/lg/ (accessed 18 Nov. 2005).

——, "Online Defamation Law," 2005, http://www.eff.org/ bloggers/lg/faq-defamation.php (accessed 18 Nov. 2005).

Erickson, Thomas, "Social Interaction on the Net: Virtual Community or Participatory Genre?" *SIGGROUP Bulletin* 18(2) (1997).

Farrell, Henry, "The Blogosphere as a Carnival of Ideas," *Chronicle of Higher Education* 52.7 (7 Oct. 2005), http://chronicle.com/free/v52/i07/07b01401.htm (accessed 27 Oct. 2005).

Fidler, Roger, *Mediamorphosis: Understanding New Media* (Thousand Oaks, CA: Pine Forge Press, 1997).

Fitzgerald, Brian, Gaye Middleton, and Anne Fitzgerald, *Jurisdiction and the Internet* (Sydney: Thomson/LBC, 2004).

Gans, Herbert J., *Democracy and the News* (New York: Oxford UP, 2003).

Gauntlett, David, *Media, Gender and Identity* (London: Routledge, 2002).

Giddens, Anthony, *Modernity and Self-Identity: Self and Society in the Late Modern Age* (Cambridge: Polity, 1991).

Gill, Kathy, "How Can We Measure the Influence of the Blogosphere?" *WWW2004 Conference Proceedings*, University of Washington, 2004.

Gillmor, Dan, *We the Media: Grassroots Journalism by the People for the People* (Sebastopol, CA: O'Reilly, 2004).

Godin, Seth, *Who's There? Seth Godin's Incomplete Guide to Blogs and the New Web*, http://sethgodin.typepad.com/seths_blog/files/whos_there.pdf (accessed 12 Sep. 2005).

Gurak, Laura J., Smiljana Antonijevic, Laurie Johnson, Clancy Ratliff, and Jessica Reyman, eds., *Into the Blogosphere: Rhetoric, Community, and Culture of Weblogs*, June 2004, http://blog.lib.umn.edu/blogosphere/ (accessed 27 Oct. 2005).

Habermas, Jürgen, *The Structural Transformation of the Public Sphere* (Cambridge, MA: MIT P, 1991).

Harrison, Teresa M., and Timothy Stephen, *Computer Networking and Scholarly Communication in the Twenty-First-Century University* (Albany: State U of New York P, 1996).

Henning, Jeffrey, "The Blogging Iceberg—Of 4.12 Million Hosted Weblogs, Most Little Seen, Quickly Abandoned," *Perseus Development Corp. White Papers* 2003, http://www.perseus.com/blogsurvey/thebloggingiceberg.html (accessed 20 Aug. 2005).

Herring, Susan C., "Slouching toward the Ordinary: Current Trends in Computer-Mediated Communication," *New Media & Society* 6.1 (2004), pp. 26–36.

——, Inna Kouper, John C. Paolillo, Lois Ann Scheidt, Michael Tyworth, Peter Welsch, Elijah Wright, and Ning Yu, "Conversations in the Blogosphere: An Analysis 'From the Bottom Up,'" *Proceedings of the 38th Annual Hawaii International Conference on Systems Sciences*, 2005.

——, Lois Ann Scheidt, Sabrina Bonus, and Elijah Wright, "Bridging the Gap: A Genre Analysis of Weblogs," *Proceedings of the 37th Hawaii International Conference on System Sciences*, 2004.

Himanen, Pekka, *The Hacker Ethic* (New York: Random House, 2001).

Hofstede, Geert, "Geert Hofstede Cultural Dimensions," 2003, http://www.geert-hofstede.com/index.shtml (accessed 2 Nov. 2004).

Israel, Shel, and Robert Scoble, *Naked Conversations* (Indianapolis: Wiley, 2006), chapter drafts available at http://redcouch.typepad.com/weblog/ (accessed 12 Sep. 2005).

Jacobs, Joanne, "The Rise of Blogs as a Product of Cybervoyeurism," *Proceedings of the Australian and New Zealand Communication Association Conference*, Gold Coast, Australia, July 2003.

Krauss, Robert M., and Susan R. Fussell, "Constructing Shared Communicative Environments," in *Perspectives on Socially Shared Cognition*, Lauren B. Resnick, John M. Levine, and Stephanie D. Teasley, eds. (Washington, DC: American Psychological Association, 1991).

Landow, George P., *Hypertext 2.0: The Convergence of Contemporary Critical Theory and Technology* (Baltimore: Johns Hopkins UP, 1997).

Lessig, Lawrence, *Code and Other Laws of Cyberspace* (New York: Basic Books, 1999).

——, *Free Culture: How Big Media Uses Technology and the Law to Lock Down Culture and Control Creativity* (New York: Penguin Press, 2004).

——, *The Future of Ideas: The Fate of the Commons in a Connected World* (New York: Random House, 2001).

Levine, Frederick, Christopher Locke, David Searles, and David Weinberger, *The Cluetrain Manifesto: The End of Business as Usual* (Cambridge, MA: Perseus, 2000).

Levy, Pierre, *Collective Intelligence: Mankind's Emerging World in Cyberspace* (Cambridge, MA: Perseus, 1997).

Li, Charlene, "Blogging: Bubble or Big Deal? When and How Businesses Should Use Blogs," 2004, http://www.forrester.com/Research/Document/Excerpt/0,7211,35000,00.html (accessed 20 Sep. 2005).

Malik, Attiya, "Are You Content with the Content? Intellectual Property Implications of Weblog Publishing," *The John Marshall Journal of Computer and Information Law* 21.439 (2003).

Meikle, Graham, *Future Active: Media Activism and the Internet* (New York: Routledge, 2002).

Miles, Adrian, "Media Rich versus Rich Media (or Why Video in a Blog Is Not the Same as a Video Blog)," *Blogtalk Downunder*, Sydney 2005, http://incsub.org/blogtalk/?page_id=74 (accessed 16 Nov. 2005).

Mortensen, Torill, and Jill Walker, "Blogging Thoughts: Personal Publication as an Online Research Tool," *Researching ICTs in Context*, Andrew Morrison, ed. (Oslo: InterMedia Report, 2002).

Nardi, Bonnie A., Diane J. Schiano, and Michelle Gumbrecht, "Blogging as a Social Activity, or, Would You Let 900 Million People Read Your Diary?" *Proceedings of the 2004 ACM Conference on Computer Supported Collaborative Work*, 2004.

Nelson, Ted, "Fixing the Computer World," *Incubation 3*, Nottingham, 2004.

——, *Literary Machines 91.1: The Report on, and of, Project Xanadu Concerning Word Processing, Electronic Publishing, Hypertext, Thinkertoys, Tomorrow's Intellectual Revolution, and Certain Other Topics Including Knowledge, Education and Freedom* (Sausalito, CA: Mindful Press, 1992).

Ojala, Marydee, "Weaving Weblogs into Knowledge Sharing and Dissemination," *Proceedings of the Nordic Conference on Information and Change*, http://www2.db.dk/NIOD/ojala.pdf, Denmark, Sep. 2004 (accessed Aug. 2005).

Oldenberg, Ray, *The Great Good Place: Cafes, Coffee Shops, Bookstores, Bars, Hair Salons, and Other Hangouts at the Heart of a Community* (New York: Marlowe, 1999).

Poster, Mark, "CyberDemocracy: Internet and the Public Sphere," Lecture at the University of California, Irvine, 1995, http://www.uoc.edu/in3/hermeneia/sala_de_lectura/mark_poster_cyberdemocracy.htm (accessed 16 Nov. 2005).

Rainie, Lee, "The State of Blogging," *Pew Internet & American Life Project*, 2 Jan. 2005, http://www.pewinternet.org/pdfs/PIP_blogging_data.pdf (accessed 5 Jan. 2005).

Rheingold, Howard, *The Virtual Community: Homesteading on the Electronic Frontier* (New York: HarperPerennial, 1994).

Rushkoff, Douglas, *Open Source Democracy: How Online Communication Is Changing Offline Politics* (London: Demos, 2003), http://www.demos.co.uk/opensourcedemocracy_pdf_media_public.aspx (accessed 22 April 2004).

Schultz, Julianne, *Reviving the Fourth Estate: Democracy, Accountability and the Media* (London: Cambridge UP, 1998).

Shirky, Clay, "A Group is its Own Worst Enemy", *Clay Shirky's Writings about the Internet: Economics & Culture, Media & Community, Open Source*, 2003, http://www.shirky.com/writings/group_enemy.html (accessed October 2003).

——, "Broadcast Institutions, Community Values," *Clay Shirky's Writings about the Internet: Economics & Culture, Media & Community, Open Source*, 9 Sep. 2002, http://www.shirky.com/writings/broadcast_and_community.html (accessed 31 May 2004).

——, "Group as User: Flaming and the Design of Social Software," *Clay Shirky's Writings about the Internet: Economics & Culture, Media & Community, Open Source*, 5 Nov. 2004, http://www.shirky.com/writings/group_user.html (accessed 21 Sep. 2005).

——, "Power Laws, Weblogs, and Inequality," *Clay Shirky's Writings about the Internet: Economics & Culture, Media & Community, Open Source*, 2 Oct. 2003, http://www.shirky.com/writings/powerlaw_weblog.html (accessed 20 Feb. 2004).

——, "RIP the Consumer, 1900-1999," *Clay Shirky's Writings about the Internet: Economics & Culture, Media & Community, Open Source*, May 2000, http://www.shirky.com/writings/consumer.html (accessed 31 May 2004).

——, "Social Software and the Politics of Groups," *Clay Shirky's Writings about the Internet: Economics & Culture, Media & Community, Open Source*, http://shirky.com/writings/group_politics.html, 9 March 2003 (accessed 17 July 2005).

——, "Weblogs and the Mass Amateurization of Publishing," *Clay Shirky's Writings about the Internet: Economics & Culture, Media & Community, Open Source*, http://www.shirky.com/writings/weblogs_publishing.html, 3 Oct. 2002 (accessed 17 July 2005).

Sifry, David, "State of the Blogosphere, August 2005, Part 5: The A-List and the Long Tail," *David Sifry's Alerts*, Aug. 2005, http://www.technorati.com/weblog/ 2005/08/39.html, 2005 (accessed 16 Nov. 2005).

Smith, Marc, and Peter Kollock, eds., *Communities in Cyberspace* (London: Routledge, 1999).

Sookman, Barry, *Computer, Internet and Electronic Commerce Law* (Toronto: Carswell Canada, 1991).

Spender, Dale, *Nattering on the Net: Women, Power and Cyberspace* (North Melbourne: Spinifex, 1995).

Tapscott, Don, *Digital Capital: Harnessing the Power of Business Webs* (Cambridge, MA: Harvard Business School P, 2000).

Van Dijck, José, "Composing the Self: Of Diaries and Lifelogs," *Fibreculture Journal* 3 (2004), http://journal.fibreculture.org/issue3/issue3_vandijck.html (accessed 27 Oct. 2005).

Warschauer, Mark, *Technology and Social Inclusion: Rethinking the Digital Divide* (Cambridge, MA: MIT P, 2003).

Weinberger, David, *Small Pieces, Loosely Joined* (London: Cambridge UP, 1998).

Wellman, Barry, and Catherine A. Haythornthwaite, eds., *The Internet in Everyday Life* (Malden, MA: Blackwell, 2002).

Williams, Jeremy B., and Joanne Jacobs, "Exploring the Use of Blogs as Learning Spaces in the Higher Education Sector," *Australasian Journal of Educational Technology* 20 (2004), pp. 232–47.

Wright, Jeremy, *Blog Marketing: The Revolutionary New Method to Increase Sales, Growth and Profits* (New York: McGraw-Hill, 2005).

General Editor: **Steve Jones**

Digital Formations is an essential source for critical, high-quality books on digital technologies and modern life. Volumes in the series break new ground by emphasizing multiple methodological and theoretical approaches to deeply probe the formation and reformation of lived experience as it is refracted through digital interaction. **Digital Formations** pushes forward our understanding of the intersections—and corresponding implications—between the digital technologies and everyday life. The series emphasizes critical studies in the context of emergent and existing digital technologies.

Other recent titles include:

Leslie Shade
  Gender and Community in the Social Construction of the Internet

John T. Waisanen
  Thinking Geometrically

Mia Consalvo & Susanna Paasonen
  Women and Everyday Uses of the Internet

Dennis Waskul
  Self-Games and Body-Play

David Myers
  The Nature of Computer Games

Robert Hassan
  The Chronoscopic Society

M. Johns, S. Chen, & G. Hall
  Online Social Research

C. Kaha Waite
  Mediation and the Communication Matrix

Jenny Sunden
  Material Virtualities

Helen Nissenbaum & Monroe Price
  Academy and the Internet

To order other books in this series please contact our Customer Service Department:
  (800) 770-LANG (within the US)
  (212) 647-7706 (outside the US)
  (212) 647-7707 FAX
To find out more about the series or browse a full list of titles, please visit our website:
  WWW.PETERLANG.COM